Arduino

LED

Projects

Robert J Davis II

Arduino LED projects by Robert J Davis II

LED stands for Light Emitting Diode. We will be using one or more LED's in every project in this book. An LED is a diode in that current only flows when power is connected in one direction. LED's usually light up at around 1.6 volts. Since our power source is 5 volts we need to add resistors to keep from damaging the LED's. Resistors resist the flow of electricity. Since an ideal current for a LED is 10 ma or .01 amps and R=E/I, (5-1.6 volts)/.01 is 3.4/.01 or 340 ohms. So the resistor that we use should be larger than 340 ohms. For the first few projects we can use 470 or 1000 ohm resistors.

The reader or builder takes all responsibility for the safe construction and operation of the projects that are contained in this book. It is assumed that the reader has at least a minimal background in programming and in electronics. It is highly recommended that you first read at least one of the two books found in the bibliography. This book does not go into as much detail as to how to write the software, or how to build the hardware.

The projects found in this book are meant to be simple enough for the average beginner. The parts counts were kept to a bare minimum to get the job done. If I have time later on I will write a second book that will go into more complex circuits and designs. The basics are given here so that with some more experimenting someone could build more complex circuits.

Each project in this book has a quick explanation, a schematic, a list of parts needed, and a software or sketch listing. This book is only meant to get you started at building some Arduino based projects. Feel free to modify, improve, or even "play" with the software and hardware. Electronics can be lots of fun and that is what you should do with these projects. Have fun!

Most of the parts to build the projects in this book are available at your local Radio Shack store. If you have lots of patience then you can also find them for sale on eBay. Be aware that many of the sellers on eBay are in located in China and hence it will take a long time for your parts to be shipped to you from China.

Chapters:

Introduction

Chapter 1

Introduction to

Electronic Components

In this book we will be using several electronic components. This is only a brief introduction to some of the components that we will be using in this book.

A "resistor" is a device that resists or limits the flow of electricity. Usually it is made out of carbon, but these days there are lots of things that are used to make resistors. This is what some resistors look like. There are one quarter watt resistors on the left, one half watt resistors on the right and a one watt resistor on the far right.

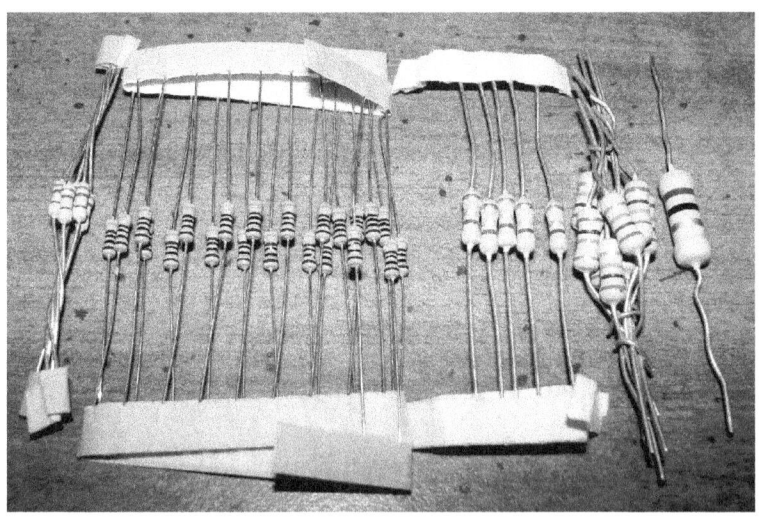

This is the schematic symbol for a resistor.

There is also what is called a "Variable resistor". These resistors can be varied by turning a knob or adjusted with a screwdriver. They can also be connected to a power source such that they can adjust the voltage out through a range of voltages.

Here are some typical variable resistors.

This is the schematic symbol for a variable resistor.

Diodes are devices that only allow current to flow in one direction. Because of that they can be used to convert AC (Alternating Current) to DC (Direct Current). They are also used as protection devices to protect delicate electronics from surges.

Up next is a picture of some typical diodes.

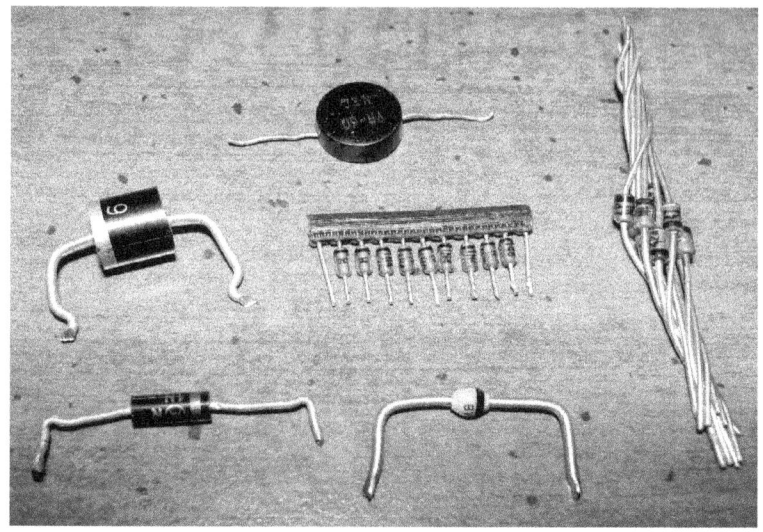

The schematic symbol for a diode is an arrow pointing in one direction.

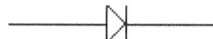

Next up we have the LED or Light Emitting Diode. This device is a diode in the electricity will only flow in one direction. When electricity flows through it, the LED will light up. Most LED's light up at around 1.6 volts. Since most sources of electricity are a lot more than 1.6 volts, a resistor should be placed in series with a LED to drop the voltage down to the correct voltage for the LED.

Here is a picture of some typical LED's, an array of LED's and several individual LED's.

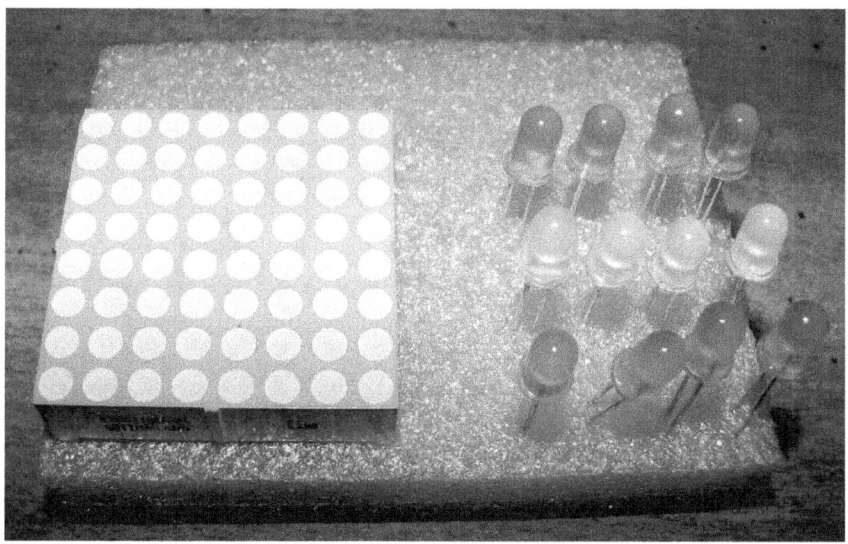

This is the schematic symbol for a LED.

Switches are devices that can turn on and turn off electricity or they can reroute it so it goes to a different place. They come in Normally Open (NO), Normally Closed (NC), Single Pole Single Throw (SPST) Double Pole Double Throw (DPDT) and many versions of Multiple Pole Multiple Throw. Some switches have as many as 12 positions or more.

Up next is a picture showing some typical switches. The top row has three momentary contact type, the bottom row has some SPST and DPDT switches.

Here are some schematic diagrams of some typical switches.

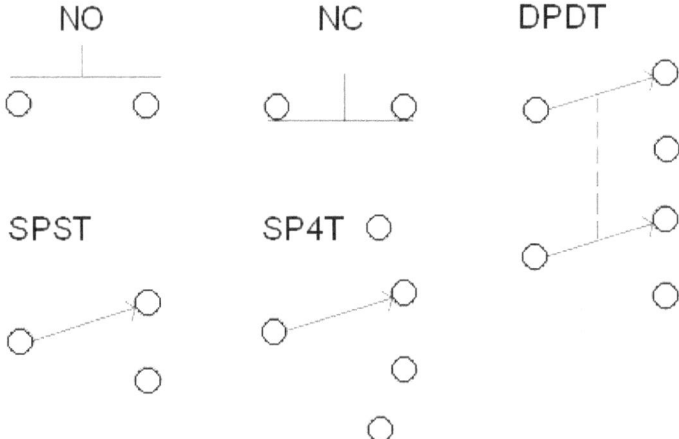

Transistors are like switches, but they work electronically. They can not only turn things on and off, but they can amplify too. That means that they can take a smaller voltage or current in and then turn on or off a larger voltage or current. Here is a picture of some typical transistors. The one on the right is a "power transistor".

Here are some schematic diagrams of typical transistors.

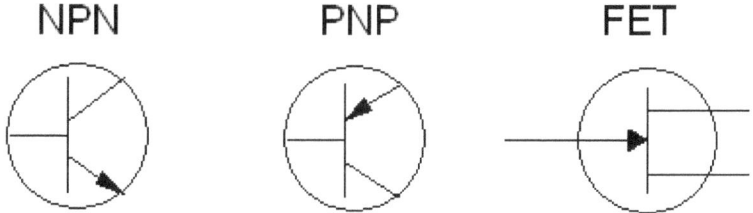

If you put many transistors together into one "case" you have what is called an "Integrated Circuit or IC for short. Some have as few as eight transistors, and some have many millions of transistors inside of them.

Here is a picture of some typical IC's.

IC's are usually diagramed as a box with a bunch of wires going in or out of it. Their connections have pin numbers that start with pin 1 at the notched end, and they then go counter clockwise around the IC when viewed from above. Here is the schematic diagram of a 74HC595 IC.

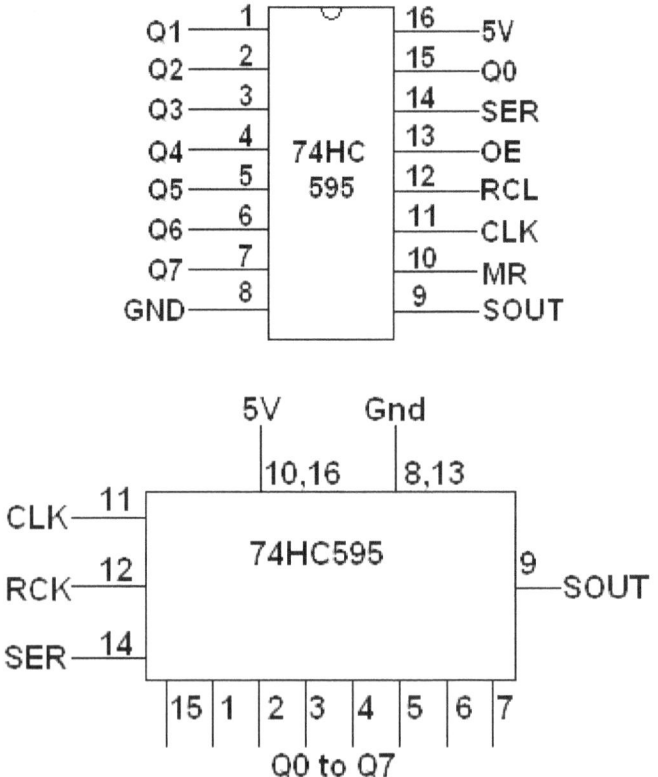

A "Breadboard" is a device that allows you to easily and temporally make and troubleshoot an electronic design. It has several connectors inside that connect sets of pins together.

Here is a picture of what a breadboard looks like with the cover removed and a row of connectors exposed. There are horizontal rows that the components are connected on, and there are vertical rows (top to bottom) that are usually used as power supply busses.

Chapter 2

Introduction to Microprocessors

Microprocessors have been around for a long time. Actually most home computers in use today are based upon microprocessors. However what we are referring to here in this book are the microprocessors that are primarily used by hobbyists.

A microprocessor runs what is called "Code" or "Software". This software works like a shopping list. In a typical shopping list you are told to go to the fruit isle and get an orange, then go to the cereal isle and get fruit loops, then go to the meat section and get some chicken wings. In much the same way a microprocessor executes commands one at a time. Microprocessors are rated on how much code they can hold in RAM (Random Access Memory) and ROM (Read Only Memory), on what I/O (Inputs and Outputs) they have, and on how fast they can run the code.

The first hobbyist microprocessor that was commonly used was the 8052AH Basic. It was an enhanced version of the 8051. There are many versions of the 8051 available today such as the 80C451. The 8052AH Basic featured serial communication with automatic baud rate detection and it included Basic in ROM. The two main drawbacks were that to add more memory you had to sacrifice several I/O Pins and that it had no analog input inputs. As a result you usually memory mapped the I/O devices such as an analog to digital converter.

The next common hobbyist microprocessor was the 68HC11. It offered 8 analog inputs, and up to six eight bit digital input and output ports depending on the version. That is a total of up to 56 I/O pins on the F1 version. It also featured built in serial communications. It had 512 bytes of EEPROM and 1024 bytes of RAM for your programs. The drawbacks were that it was programmed in assembler, and that you had to program it on one circuit board and then transfer the CPU to the "live" circuit board or sacrifice a lot of I/O pins. There were many versions of the 68HC11 that were used in many different applications. You might even have one that is built into and that is actually running your car!

This picture is of a typical home made 68HC11-F1 controller. The eight digital inputs are on the left, the eight analog inputs are across the top, and the eight digital outputs are on the right. There is a ULN2008 driver on the digital outputs so they can drive output relays. The voltage regulator is located in bottom center and the RS-422 serial is located on the bottom left. There are four DIP (Dual Inline Package) switches on the bottom right to select the serial address so that several of these controllers can be connected together on the same RS-422 serial buss.

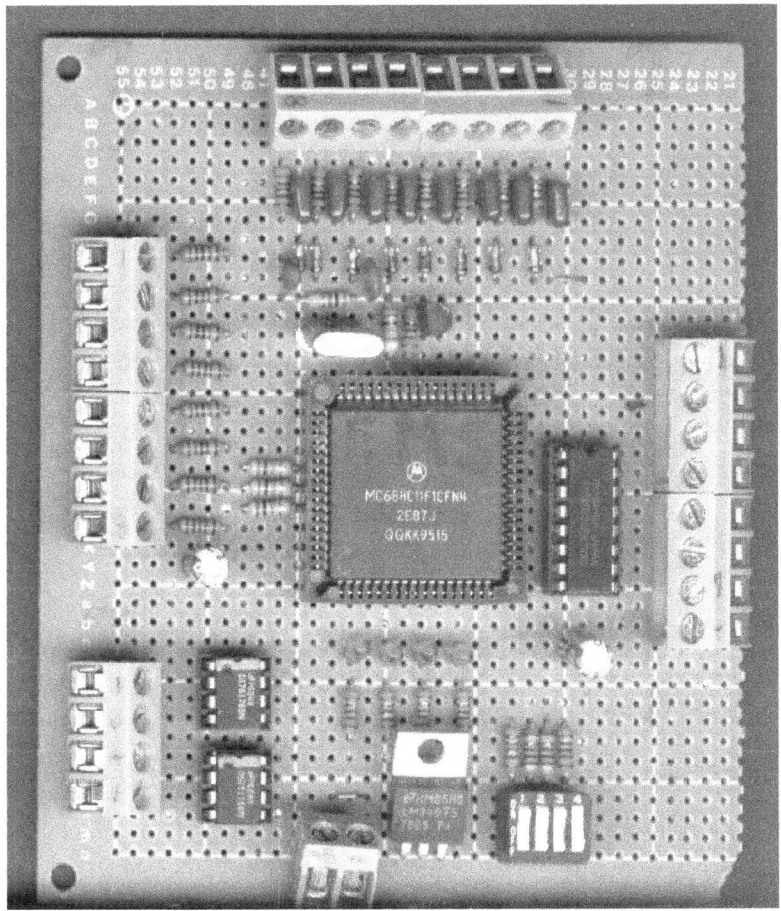

The next common hobbyist microprocessor was called the "Basic Stamp". It featured Basic in ROM. The CPU and all the required support parts were mounted on a small 24 pin chip about the size of a large postage stamp. It featured 2048 bytes of EEPROM and 32 bytes of RAM for your programs. It featured 16 I/O pins. Drawbacks of the Basic Stamp included that there were no analog inputs and Basic is relatively slow. However this was the most popular hobby microprocessor up until very recently.

In the picture below, the Basic Stamp chip is the red chip that is top and center. On the left side is the serial interface jack, and on the right side is the optional voltage regulator. In the bottom breadboard area there are two seven segment LED displays that were used for a Basic Stamp project.

The latest common hobbyist microprocessor, and the subject of this book, is called the "Arduino". It features both six analog inputs, 14 digital I/O pins and a higher level programming language. It is essentially programmed in a version of C. It also offers 14 digital, software selected, input or output pins. Some of the I/O pins can be configured as analog outputs as well. It has 1024 bytes of EEPROM, 2028 bytes of RAM and then a generous 32K (that's 32,000 bytes) of Flash memory for your applications. One of the best features is the support for programming the Arduino via a USB cable. Many newer computers are lacking the serial ports that were needed to work with most of the older microprocessors.

The picture below is of an Arduino UNO revision 3. On the left side there are eight power pins with reset, regulated, 3.3 volts, regulated 5 volts, ground, and unregulated "Vin" (Voltage input from the power jack) outputs. Below that on the left side there are the six analog input pins. On the right side there are 14 digital input or output pins.

There is also a ground and Analog to Digital (A to D) converter reference voltage output available on the right side. Across the top, on the left there is a power jack for 6 to 9 volts DC from an optional AC adapter and on the right there is a USB jack to plug into a computer to program the Arduino microprocessor.

Below is a picture of an Arduino clone. It is a clone of an earlier version so it has less pins. The labels on the pins are much easier to read. Notice that overall it is very similar to the Arduino above.

16

Chapter 3

Setting up an Arduino

To get started with the Arduino it is recommended that you first read one or two of the books that are found in the bibliography. However here is a very brief set of instructions that might actually be sufficient to get you started with an Arduino.

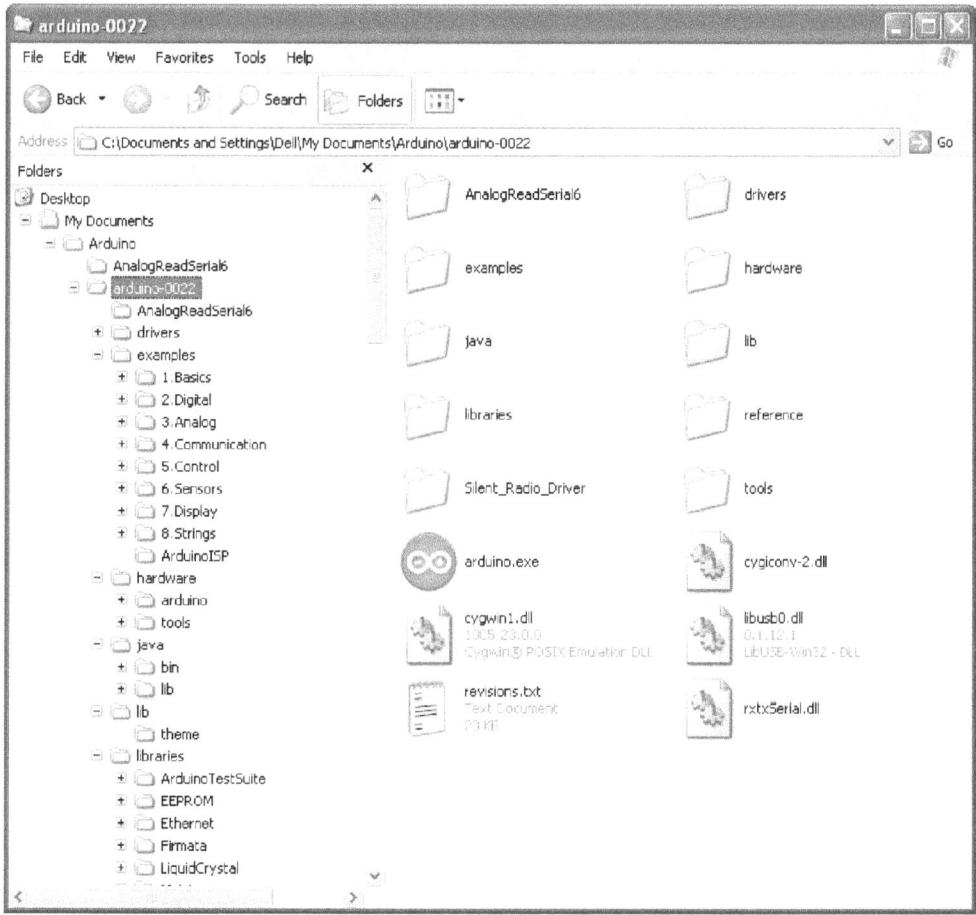

First you will need to download and then unzip the Arduino driver software. In the picture above the driver software package is called "Arduino-0022" it is shown unzipped in a folder called "Arduino".

Unfortunately when you plug an Arduino into the USB port it is not automatically recognized by Windows. You have to steer the Widows hardware installer to the directory where you have unzipped the drivers so that it can find the DLL files as seen in the picture below.

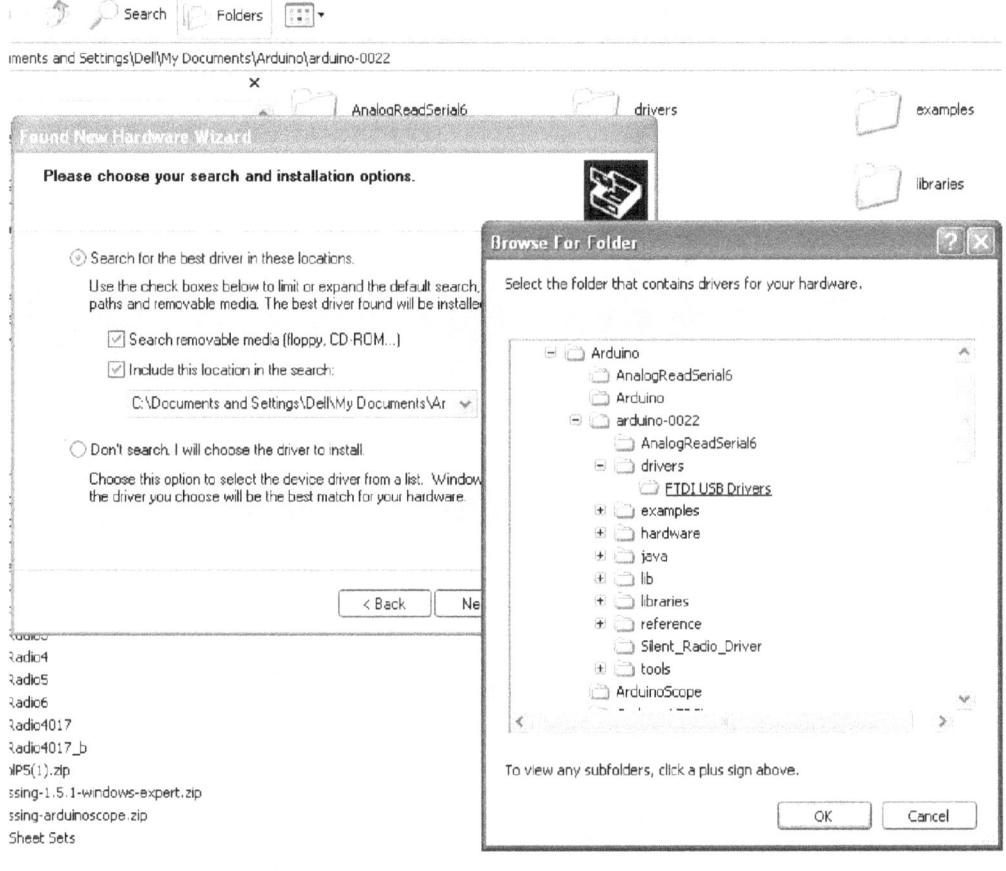

Also note the "Arduino.exe" file that was listed among the unzipped files two pictures back. It is what you will need to click on to start working with the Arduino after you plug in the serial cable and get the drivers properly installed. I usually create a shortcut to Arduino.exe on the desktop of my computer. When you click on Arduino.exe you should see this:

You can now cut and paste your code into the white area below the "sketch" tab. To upload your code into the Arduino you will need to make sure what your model of Arduino is and that the serial ports are properly selected first. They are set up under "tools" "board" and "tools" "serial port". I have sometimes had to guess at the board name on some Arduino clones.

Below is the list of the many supported Arduino boards at this time. Note that they are mostly based on either the ATMega168 or the ATMega328 chips. The second question is does it operate on 5 volts or on 3.3 volts? The projects in this book assume 5 volt operation but they likely will also work on 3.3 volt versions of the Arduino as well.

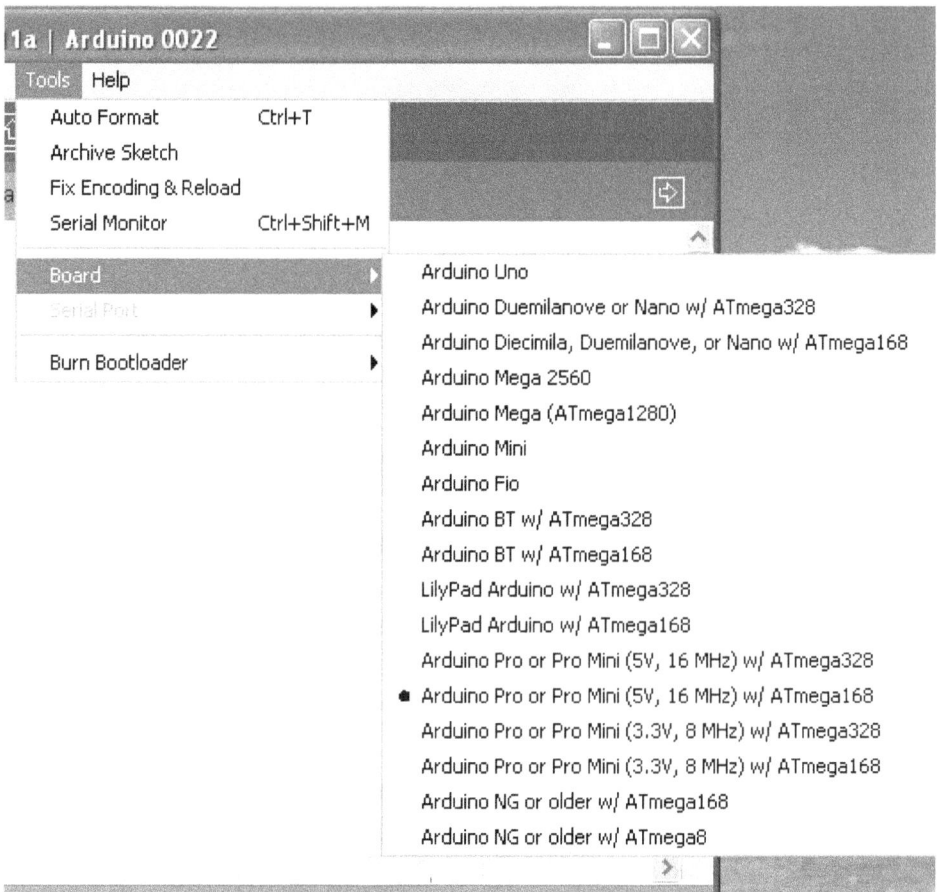

If everything seems to be working properly, you can now connect a couple of LED's to the Arduino. Then you can upload some simple code to see if the LED's blink. Hopefully you are now up and running so you can start building the circuits found in this book.

Every Arduino program or "Sketch" should have four parts. First you should identify the program, the author, the date, and perhaps the version. It could look something like this:

```
//  Name    : LED back and forth blinker
//  Author  : Bob Davis
//  Date    : 5 March, 2013
//  Version : 1.0
```

Comments that are to be ignored by the Arduino can start with "//".
Secondly you need to establish your variables, like this. Note that "int"
stands for integer.

```
// Pins for the LED's
int LED1Pin = 1;
int LED2Pin = 2;
```

Thirdly you need to have what is called "setup( )". It establishes things
that only need to be done only once. Since the digital pins can be either
input or output pins you will need to tell the Arduino if they will be used
for input or output.

```
// Set the pins to output to the LED's
void setup() {
  pinMode(LED1Pin, OUTPUT);
  pinMode(LED2Pin, OUTPUT);
}
```

Lastly you have the heart of the program, it is called the "loop( )". This
simple program will blink the two LED's back and forth, when they are
connected to Arduino digital I/O pins one and two.

```
void loop() {
   digitalWrite (LED1Pin, HIGH);
   digitalWrite (LED2Pin, LOW);
 delay(500);
   digitalWrite (LED2Pin, HIGH);
   digitalWrite (LED1Pin, LOW);
 delay(500);
 }
```

Most of the circuits in this book were first made on what is called a
"breadboard". Here is a picture of a typical breadboard setup with an
Arduino clone. The blue wire is connected to ground. If the LED's are
connected backwards they will not light up. There is a flat spot on the
LED's right or ground side in the picture.

In the picture above the Arduino clone is on the left and on the right is a small breadboard. You can use short jumper wires from the Arduino outputs to the breadboard. That arrangement would make the LED's easier to set up. The LED's light up at 1.6 volts so resistors need to be wired in series so that they do not overheat or damage the five volt Arduino outputs or the LED's. I used 470 ohm resistors, but any value from 100 ohms all the way up to 1000 ohms should work just fine.

The input switches that are used in some of these projects should not require the use of pull up resistors to work properly. There are internal pull up resistors inside of the Arduino processor. They are turned on by writing a "HIGH" to an input pin such as: "digitalWrite (inputpin, HIGH)".

Here is a picture of typical input switches with wires soldered on to be able to plug them into a breadboard or the Arduino.

There are some limits on what you can do with what I/O pin. For instance D0 has to be an input pin. If you make it an output pin you will likely loose the ability to upload code to the Arduino via USB. I am speaking form experience here. Also D1 should be an output pin. If it is used as an input, and the input is ground, the Arduino will not be able to respond to the USB data. Pin D13 should be an output pin. If D13 is used as an input the built in pull up resistor is overridden by a LED that is permanently installed on the circuit board.

The analog input pins can be called A0 to A5 or they can be called D14 to D19. Besides that they can also be used as digital input or as digital output pins. To change the analog pins mode just use the "pinMode(A0, OUTPUT)" or "pinMode(A0, INPUT)" command. Once it is a digital input pin, you can also activate the internal pull up resistor.

Chapter 4

Electronic Coin Flipper

Most Arduino project books will start you off with a project that blinks one or two LED's. Here we will start off with two LED's and then add a switch to make something that might actually be useful. It is an "Electronic Coin Flipper". Each time you push the switch, it will randomly come up with either the "Heads" or "Tails" LED's.

It works by reading the analog input pin to see if the switch is being depressed before advancing to the next LED. As long as the switch is held down the LED's will rapidly blink back and forth. When you stop pushing down on the switch the LED's will stop at the last one.

Here is the wiring schematic:

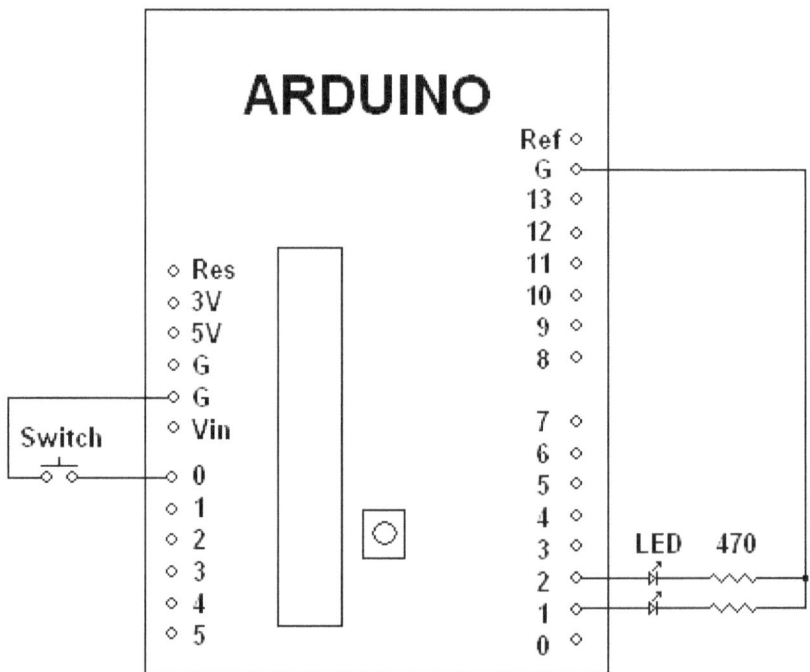

Here is a list of the needed parts to make this work, besides the Arduino itself.

1. Normally open, momentary contact switch ½ inch in diameter.
2. Two LED's that that can be any color.
3. Two 470 ohm resistors

Here is the code or sketch to make it work. Note that I switched the analog inputs into digital input mode and turned on the internal pull up resistor.

```
//*************************************************//
//  Name    : LED Driver Coin Flip
//  Author  : Bob Davis
//  Date    : 5 March, 2013
//  Version : 1.1
//*************************************************//
// Pins for the LED's
int LED1Pin = 1;
int LED2Pin = 2;
int LED3Pin = 3;
int LED4Pin = 4;
int LED5Pin = 5;
int LED6Pin = 6;
int LED7Pin = 7;
// Set the pins to output to the LED's
void setup() {
  pinMode(LED1Pin, OUTPUT);
  pinMode(LED2Pin, OUTPUT);
  pinMode(LED3Pin, OUTPUT);
  pinMode(LED4Pin, OUTPUT);
  pinMode(LED5Pin, OUTPUT);
  pinMode(LED6Pin, OUTPUT);
  pinMode(LED7Pin, OUTPUT);
  pinMode(A0, INPUT);
  digitalWrite (A0, HIGH);
}
void loop() {
  if (digitalRead(A0) == LOW){
    digitalWrite (LED1Pin, HIGH);
    digitalWrite (LED2Pin, LOW);
```

```
}
delay(50);
if (digitalRead(A0) == LOW){
  digitalWrite (LED2Pin, HIGH);
  digitalWrite (LED1Pin, LOW);
}
delay(50);
}
```

Chapter 5

Electronic Dice

If we take the electronic coin flipper one step further we can make electronic dice. We will need to arrange the LED's so that they will appear like the dots on the faces of dice. We will also have to make the software light up the numbers for one to six in the correct arrangements to match what you would find on the face of the dice.

Here once again, when you hold down the switch, the counter advances and when you let off the switch it stops at whatever number was last. However if you look closely the software the counter never actually stops running! That way the next time that you push the switch it does not continue where it left off, but comes up "randomly".

Here is the arrangement of the LED's

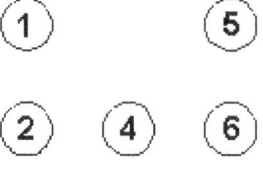

This would be the order that the LED's would light up in:

1   4
2   2, 6
3   2, 4, 6
4   1, 3, 5, 7
5   1, 3, 4, 5, 7

| 6 | 1, 2, 3, 5, 6, 7 |

Here is a list of the parts that are needed to make this work, besides the Arduino itself.

1. One Normally open momentary contact switch ½ inch in diameter.
2. Seven LED's can be any color.
3. Seven 470 ohm resistors

Here is the schematic diagram of how to connect everything up:

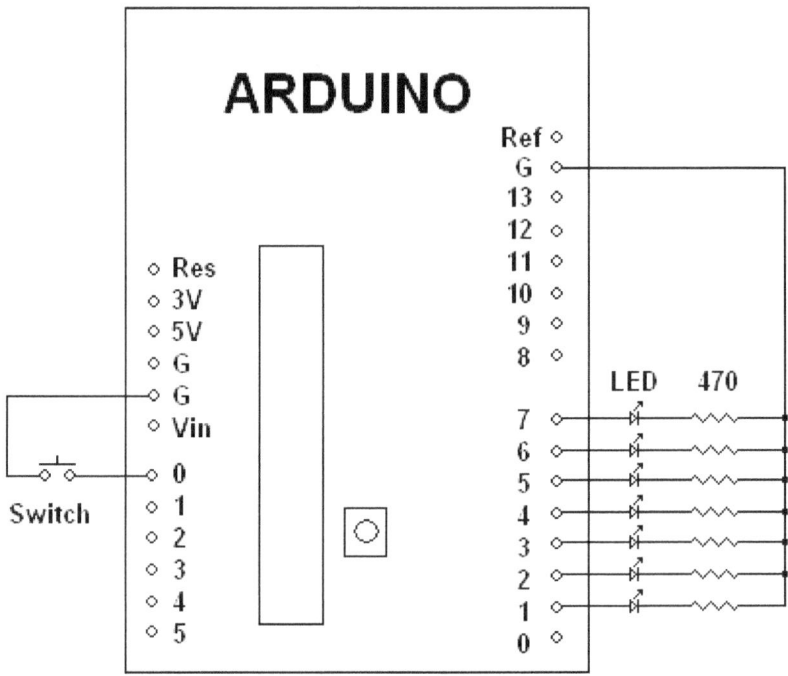

Here is the sketch or code to make it work:

```
//*************************************************//
//  Name    : LED Dice Driver
//  Author  : Bob Davis
//  Date    : 5 March, 2013
//  Version : 1.1
//*************************************************//
// Pins for the LED's
int LED1Pin = 1;
int LED2Pin = 2;
int LED3Pin = 3;
```

```
int LED4Pin = 4;
int LED5Pin = 5;
int LED6Pin = 6;
int LED7Pin = 7;
// Set the pins to output to the LED's
void setup() {
  pinMode(LED1Pin, OUTPUT);
  pinMode(LED2Pin, OUTPUT);
  pinMode(LED3Pin, OUTPUT);
  pinMode(LED4Pin, OUTPUT);
  pinMode(LED5Pin, OUTPUT);
  pinMode(LED6Pin, OUTPUT);
  pinMode(LED7Pin, OUTPUT);
  pinMode(A0, INPUT);
  digitalWrite (A0, HIGH); }
void loop() {
  for (int dice = 1; dice < 8; dice++){
  if (digitalRead(A0) == LOW){
    digitalWrite (LED1Pin, LOW);
    digitalWrite (LED2Pin, LOW);
    digitalWrite (LED3Pin, LOW);
    digitalWrite (LED4Pin, LOW);
    digitalWrite (LED5Pin, LOW);
    digitalWrite (LED6Pin, LOW);
    digitalWrite (LED7Pin, LOW);
    if (dice==1) digitalWrite (LED4Pin, HIGH);
    if (dice==2) {
      digitalWrite (LED2Pin, HIGH);
      digitalWrite (LED6Pin, HIGH); }
    if (dice==3) {
      digitalWrite (LED2Pin, HIGH);
      digitalWrite (LED4Pin, HIGH);
      digitalWrite (LED6Pin, HIGH); }
    if (dice==4) {
      digitalWrite (LED1Pin, HIGH);
      digitalWrite (LED3Pin, HIGH);
      digitalWrite (LED5Pin, HIGH);
      digitalWrite (LED7Pin, HIGH); }
    if (dice==5) {
      digitalWrite (LED1Pin, HIGH);
      digitalWrite (LED3Pin, HIGH);
```

```
    digitalWrite (LED4Pin, HIGH);
    digitalWrite (LED5Pin, HIGH);
    digitalWrite (LED7Pin, HIGH); }
  if (dice==6) {
    digitalWrite (LED1Pin, HIGH);
    digitalWrite (LED2Pin, HIGH);
    digitalWrite (LED3Pin, HIGH);
    digitalWrite (LED5Pin, HIGH);
    digitalWrite (LED6Pin, HIGH);
    digitalWrite (LED7Pin, HIGH); }
  if (dice==7) {
    digitalWrite (LED1Pin, HIGH);
    digitalWrite (LED2Pin, HIGH);
    digitalWrite (LED3Pin, HIGH);
    digitalWrite (LED4Pin, HIGH);
    digitalWrite (LED5Pin, HIGH);
    digitalWrite (LED6Pin, HIGH);
    digitalWrite (LED7Pin, HIGH); }
  }
delay(50);
} }
```

Chapter 6

Wheel of Prizes

This can physically be a very large project to build. I have used a 4 foot by 4 foot piece of plywood to make it. It could be made from a 2 or a 3 feet square piece of wood as well. The frame is made out of some 2x4's with a grove cut into them that is wide enough to fit the edge of the plywood. Basically you mount 8 bright LED's in sockets at equal intervals around a circle. In the middle of the circle there are prize panels that are either glued on or held in place by some Velcro. The operator hits the start button and the contestant then hits the stop button. Whatever prize the light stops on, is what the contestant wins.

This time we are going to use two push buttons, one to start it and one to stop it. You could easily modify the hardware design and software to have as many as 10 or even 12 prize positions around the circle if you wanted to offer more prizes.

Up next is the schematic diagram showing how to wire up the "wheel of prizes".

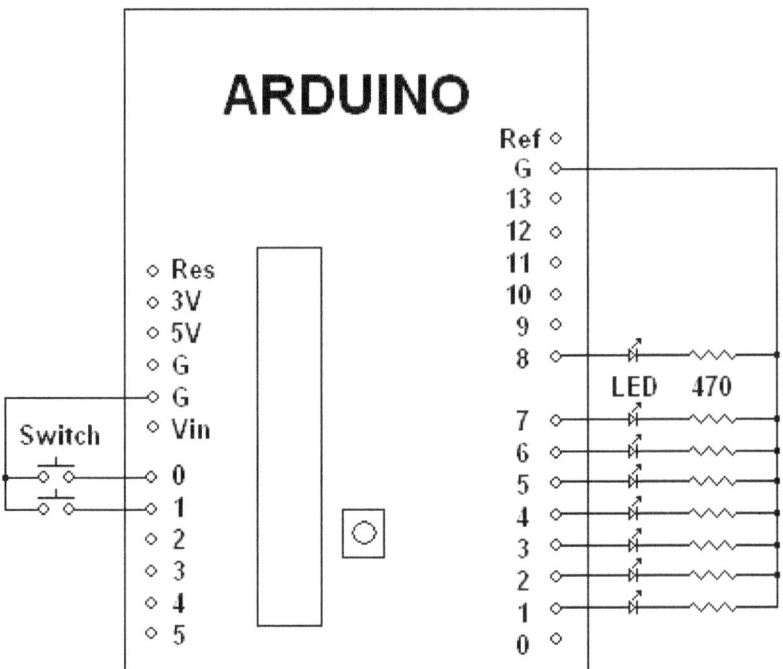

Here is a list of the needed parts to make this work, besides the Arduino itself.

1. Two normally open momentary contact switch ½ inch in diameter.
2. Eight LED's can be any color.
3. Eight 470 ohm resistors

Next is a conceptual drawing of what the completed project might look like.

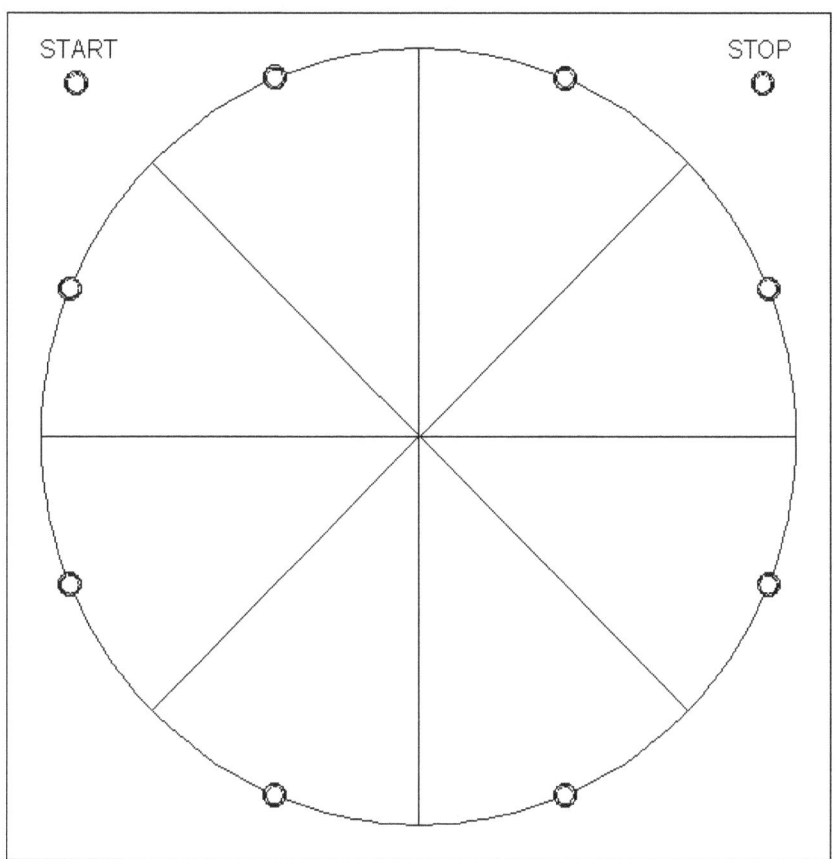

Here is the "Sketch" code to make it work:

```
//************************************************/
// Name    : LED Wheel of Prizes
// Author  : Bob Davis
// Date    : 5 March, 2013
// Version : 1.1
//************************************************/
// Pins for the LED's
int LED1Pin = 1;
int LED2Pin = 2;
int LED3Pin = 3;
int LED4Pin = 4;
int LED5Pin = 5;
int LED6Pin = 6;
int LED7Pin = 7;
int LED8Pin = 8;
```

```
int spin = 1;
int prize = 1;
// Set the pins to output to the LED's
void setup() {
  pinMode(LED1Pin, OUTPUT);
  pinMode(LED2Pin, OUTPUT);
  pinMode(LED3Pin, OUTPUT);
  pinMode(LED4Pin, OUTPUT);
  pinMode(LED5Pin, OUTPUT);
  pinMode(LED6Pin, OUTPUT);
  pinMode(LED7Pin, OUTPUT);
  pinMode(LED8Pin, OUTPUT);
  pinMode(A0, INPUT);
  pinMode(A1, INPUT);
  digitalWrite(A0, HIGH);
  digitalWrite(A1, HIGH);
}
void loop() {
  if (prize > 8) prize = 0;
  if (digitalRead(A0) == LOW) spin=1;
  if (digitalRead(A1) == LOW) spin=0;
  if (spin==1) {
    prize++;
    digitalWrite (LED1Pin, LOW);
    digitalWrite (LED2Pin, LOW);
    digitalWrite (LED3Pin, LOW);
    digitalWrite (LED4Pin, LOW);
    digitalWrite (LED5Pin, LOW);
    digitalWrite (LED6Pin, LOW);
    digitalWrite (LED7Pin, LOW);
    digitalWrite (LED8Pin, LOW);
    if (prize==1) digitalWrite (LED1Pin, HIGH);
    if (prize==2) digitalWrite (LED2Pin, HIGH);
    if (prize==3) digitalWrite (LED3Pin, HIGH);
    if (prize==4) digitalWrite (LED4Pin, HIGH);
    if (prize==5) digitalWrite (LED5Pin, HIGH);
    if (prize==6) digitalWrite (LED6Pin, HIGH);
    if (prize==7) digitalWrite (LED7Pin, HIGH);
    if (prize==8) digitalWrite (LED8Pin, HIGH);
  }
  delay(50);   }
```

Chapter 7

Back and Forth Scanner

You can make several LED's blink in succession down a line without using a microprocessor, but can you make them reverse their direction and go back and forth without using a microprocessor? I doubt it. You could add this circuit to your car and pretend that it is a computerized car, but actually it is computerized!

Here is a drawing of the idea showing how it would work. First the lit LED's travel in one direction and then they reverse and travel in the opposite direction.

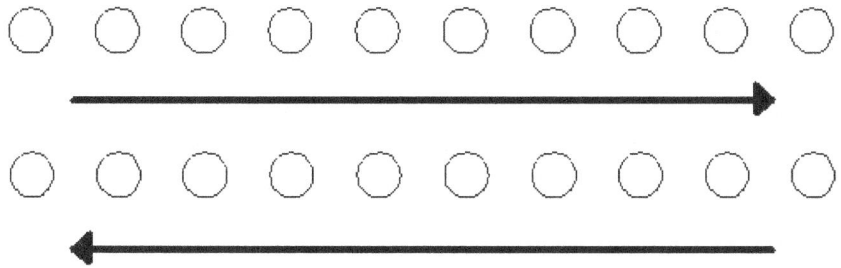

This is a picture of what one of these scanners looks like. This was first used in a "time machine". The switch on the right was an on/off switch.

Below is the wiring diagram of how to connect the LED's to the Arduino:

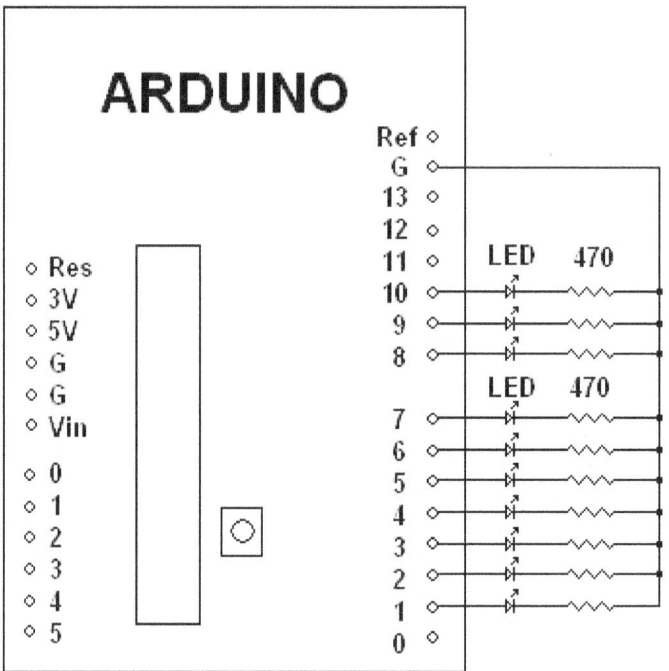

Here is a list of the parts list to make this work, besides the Arduino itself.

1. Ten LED's can be any color.
2. Ten 470 ohm resistors

Here is the software for the Arduino to make it work:

```
//*********************************************/
//  Name    : LED Back and Forth
//  Author  : Bob Davis
//  Date    : 5 March, 2013
//  Version : 1.0
//*********************************************/
// Pins for the LED's
int LED1Pin = 1;
int LED2Pin = 2;
int LED3Pin = 3;
int LED4Pin = 4;
int LED5Pin = 5;
int LED6Pin = 6;
int LED7Pin = 7;
```

```
int LED8Pin = 8;
int LED9Pin = 9;
int LED10Pin = 10;
// Set the pins to output to the LED's
void setup() {
  pinMode(LED1Pin, OUTPUT);
  pinMode(LED2Pin, OUTPUT);
  pinMode(LED3Pin, OUTPUT);
  pinMode(LED4Pin, OUTPUT);
  pinMode(LED5Pin, OUTPUT);
  pinMode(LED6Pin, OUTPUT);
  pinMode(LED7Pin, OUTPUT);
  pinMode(LED8Pin, OUTPUT);
  pinMode(LED9Pin, OUTPUT);
  pinMode(LED10Pin, OUTPUT);
}
void loop() {
  for (int scan = 1; scan < 19; scan++){
    digitalWrite (LED1Pin, LOW);
    digitalWrite (LED2Pin, LOW);
    digitalWrite (LED3Pin, LOW);
    digitalWrite (LED4Pin, LOW);
    digitalWrite (LED5Pin, LOW);
    digitalWrite (LED6Pin, LOW);
    digitalWrite (LED7Pin, LOW);
    digitalWrite (LED8Pin, LOW);
    digitalWrite (LED9Pin, LOW);
    digitalWrite (LED10Pin, LOW);
    if (scan ==1) digitalWrite (LED1Pin, HIGH);
    if (scan ==2) digitalWrite (LED2Pin, HIGH);
    if (scan ==3) digitalWrite (LED3Pin, HIGH);
    if (scan ==4) digitalWrite (LED4Pin, HIGH);
    if (scan ==5) digitalWrite (LED5Pin, HIGH);
    if (scan ==6) digitalWrite (LED6Pin, HIGH);
    if (scan ==7) digitalWrite (LED7Pin, HIGH);
    if (scan ==8) digitalWrite (LED8Pin, HIGH);
    if (scan ==9) digitalWrite (LED9Pin, HIGH);
    if (scan ==10) digitalWrite (LED10Pin, HIGH);
    if (scan ==11) digitalWrite (LED9Pin, HIGH);
    if (scan ==12) digitalWrite (LED8Pin, HIGH);
    if (scan ==13) digitalWrite (LED7Pin, HIGH);
```

```
    if (scan ==14) digitalWrite (LED6Pin, HIGH);
    if (scan ==15) digitalWrite (LED5Pin, HIGH);
    if (scan ==16) digitalWrite (LED4Pin, HIGH);
    if (scan ==17) digitalWrite (LED3Pin, HIGH);
    if (scan ==18) digitalWrite (LED2Pin, HIGH);
  delay(500);
  }
}
```

Chapter 8

Quiz Practice Machine

The Quiz practice machine is for practicing for what are called "Quiz rallies". There can be up to six contestants at one time using the Arduino analog input pins. Each contestant has a handheld switch to push if they know what the answer is. The first contestant to push his button will have his LED light come on and an optional buzzer will sound. At this point all of the other contestants are locked out. The moderator needs to push a "reset" button in order for the game to continue.

This is a picture showing what a Quiz Practice Machine looks like. Across the front (the far side) are the six LED's, three red and three green in color. They are mounted on a small circuit board. In the middle is the Arduino processor. Across the back are the power and reset connector, the two sets of three wires from the push buttons and the buzzer is on the right.

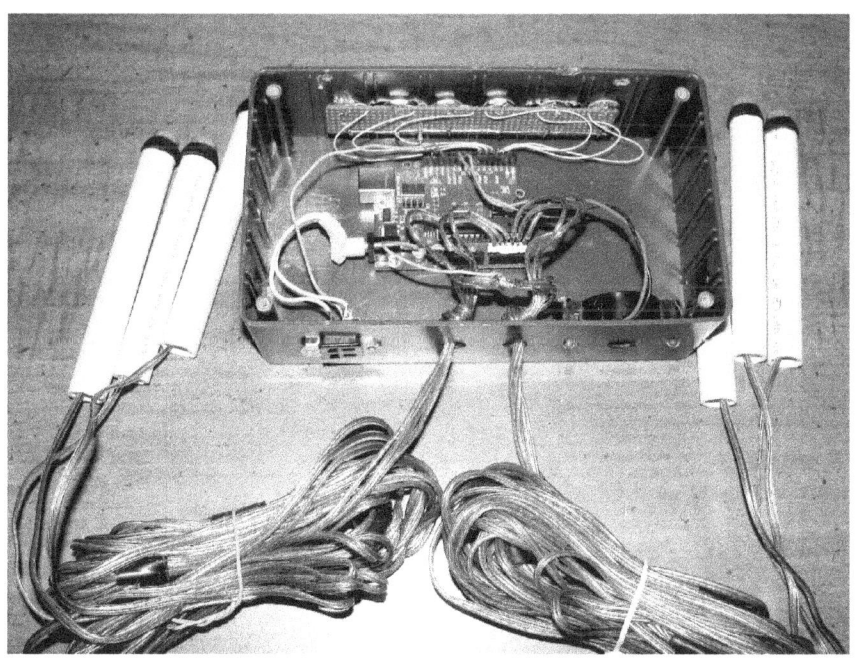

The hand held buttons are made from four to five inch long pieces of half inch diameter PVC pipe. An alternative to the PVC pipe is to use the center out of old CD spindles. A four or five foot piece of two conductor speaker wire is connected to a switch that fits neatly inside of the PVC pipe. Glue, heat shrink insulation, or electrical tape can be used to hold the switch tightly in place. The other end of the speaker wire goes to the Arduino analog input pins. Here is a picture of a typical input switch.

This is the schematic diagram, note the extra "reset" switch that is located on the right.

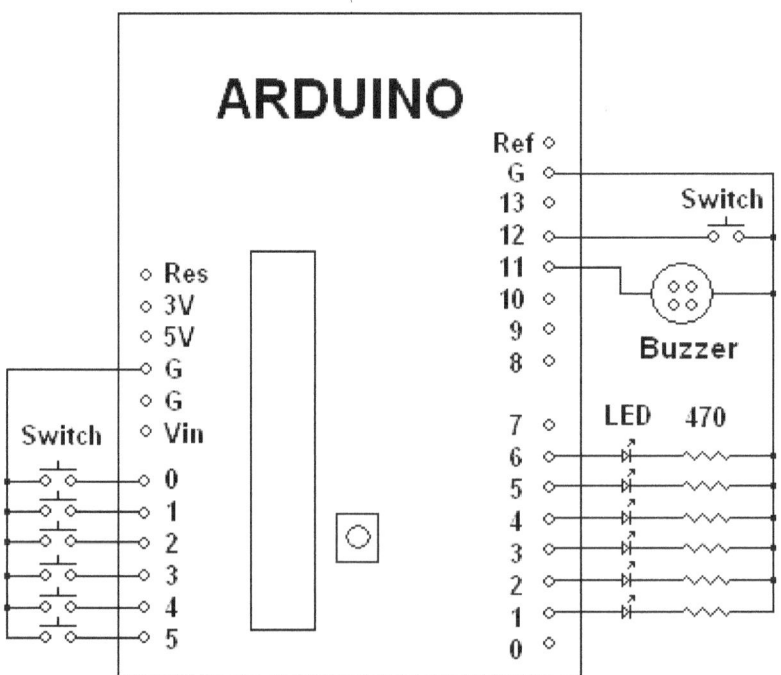

Here is a list of the needed parts to make this work, besides the Arduino itself.

1. Six normally open momentary contact switches about ½ inch in diameter.
2. One normally open momentary contact reset switch
3. Six LED's can be any color.
4. Six 470 ohm resistors.
5. Optional buzzer on pin 11

This is the software or sketch to make it work. Not that I turned the analog inputs into digital inputs. That change gave much more reliable results. I think that the analog inputs ignore the built in pull up resistors.

```
//**********************************************/
// Name    : LED Quiz Practice
// Author  : Bob Davis
// Date    : 5 March, 2013
// Version : 1.1
//**********************************************/
// Pins for the LED's
int LED1Pin = 1;
int LED2Pin = 2;
int LED3Pin = 3;
int LED4Pin = 4;
int LED5Pin = 5;
int LED6Pin = 6;
int LED7Pin = 7;
// Analog input pins A0-A5
int in0Pin = 14;
int in1Pin = 15;
int in2Pin = 16;
int In3Pin = 17;
int in4Pin = 18;
int in5Pin = 19;
int BeepPin = 11;
int ResSwitch = 12;
int latch = 0;

// Set the pins to output to the LED's
void setup() {
```

```
  pinMode(LED1Pin, OUTPUT);
  pinMode(LED2Pin, OUTPUT);
  pinMode(LED3Pin, OUTPUT);
  pinMode(LED4Pin, OUTPUT);
  pinMode(LED5Pin, OUTPUT);
  pinMode(LED6Pin, OUTPUT);
  pinMode(LED7Pin, OUTPUT);
  pinMode(in0Pin, INPUT);
  pinMode(in1Pin, INPUT);
  pinMode(in2Pin, INPUT);
  pinMode(in3Pin, INPUT);
  pinMode(in4Pin, INPUT);
  pinMode(in5Pin, INPUT);
  pinMode(ResSwitch, INPUT);
  pinMode(BeepPin, OUTPUT);
  digitalWrite(in0Pin, HIGH);
  digitalWrite(in1Pin, HIGH);
  digitalWrite(in2Pin, HIGH);
  digitalWrite(in3Pin, HIGH);
  digitalWrite(in4Pin, HIGH);
  digitalWrite(in5Pin, HIGH);
  digitalWrite(ResSwitch, HIGH);
}
void loop() {
  if (latch == 0){
    digitalWrite (LED1Pin, LOW);
    digitalWrite (LED2Pin, LOW);
    digitalWrite (LED3Pin, LOW);
    digitalWrite (LED4Pin, LOW);
    digitalWrite (LED5Pin, LOW);
    digitalWrite (LED6Pin, LOW);
    if (digitalRead(in0Pin) == LOW) {
      digitalWrite(LED6Pin, HIGH);
      latch = 1; }
    if (digitalRead(in1Pin) == LOW) {
      digitalWrite(LED5Pin, HIGH);
      latch = 1; }
    if (digitalRead(in2Pin) == LOW) {
      digitalWrite(LED4Pin, HIGH);
      latch = 1; }
    if (digitalRead(in3Pin) == LOW) {
```

```
      digitalWrite(LED3Pin, HIGH);
      latch = 1; }
    if (digitalRead(in4Pin) == LOW) {
      digitalWrite(LED2Pin, HIGH);
      latch = 1; }
    if (digitalRead(in5Pin) == LOW) {
      digitalWrite(LED1Pin, HIGH);
      latch = 1; }
  }
  if (latch == 1){
    digitalWrite (BeepPin, HIGH);
    delay(50);
    if (digitalRead(ResSwitch) == LOW){
      latch = 0;
      digitalWrite (BeepPin, LOW);  }
  }
  digitalWrite(LED7Pin, HIGH);  //power on indicator
}
```

Chapter 9

Introducing LED Arrays

When you need to use more than just a hand full of LED's it is possible to get them preassembled and prearranged into several different patterns. The first common arrangement is what is called the "seven segment" numeric display. They display the numbers zero to nine using seven LED's. These displays are used a lot in panel meters and in some older calculators. They are also available with one, two, three and I have seen as many as six digits. Below is a picture of some types of seven segment LED displays.

You will need to know that almost any LED array can be found in both "Common Cathode" (CC) and "Common Anode" (CA) arrangements. That is to say that the common side of the LED's can be grounded to work in the common cathode arrangement. The common side can be connected to power in the case of the common anode arrangement. The terms "cathode" and "anode" are from the old vacuum tube days when the cathode of a tube emitted the electrons and the anode collected the electrons. Below are two schematic diagrams showing both the common cathode and common anode LED arrangements.

Common
Cathode
(Ground)

Common
Anode
(Power)

LED

LED

If you do not know the pin connections of an LED display, it can be determined by using a nine volt battery and a 1000 ohm resistor. Just start in one corner with applying power then use the resistor to apply ground to all of the other pins. Take notes on what connections will cause what LED's to light up. With this trial and error method you can quickly reconstruct the wiring of almost any LED array. Another solution is to Google the part number and look for the wiring diagram.

Another important consideration is the drive capacity of most logic circuits. Most logic circuits can only drive about 20 ma, or .02 amps to ground. That is enough power to light one or two LED's. However the common connection to the LED's in a common cathode or common anode arrangement must power seven or eight LED's. That would take from 70 to 80 ma or .08 amps to light all seven or eight of them. This amount of power will require special driver IC's, or the use of power transistors. For less than one half of an amp you can use a 2N2222 (NPN) or a 2N2907 (PNP) transistor. For more amps, Power Darlington transistors like the TIP120 (NPN) and TIP125 (PNP) are the most popular drivers. If special drivers are not used, then the brightness of the LED's will decrease as the number of LED's that are lit increases. The Arduino can sink or source about 40 ma. or enough power to light about 4 LED's per pin.

There are the many LED matrix type of arrangements. The most common LED Matrix arrangements are five by seven, five by eight, and eight by eight. Below is a picture of some typical LED arrays. The left most array is an eight by five array, the other two are eight by eight arrays. The arrays are also in some of the more common sizes, 2 and 3/8 inches for the larger ones and 1 and 1/2 inches for the smaller ones. The smaller one on the right is an eight by eight dual color array like what we will be using for some projects in this book.

Led arrays can also be found in many colors and many combinations of colors. The most common colors are red and green but some LED arrays are available in red, green and blue so they can produce all of the colors of the rainbow by combining those three colors. By using three color LED's you can make a "TV" like LED sign. LED Signs like that are popping up all over the place.

Chapter 10

Seven Segment Display Counter

In this chapter we will be adding a seven segment display. It is essentially seven LED's prearranged so that it can show the numbers zero to nine using seven "segments". The Arduino can directly drive a seven segment LED display but if you do that, you will use seven pins per display. When you try to add a second seven segment display you will need at least 14 I/O pins. As you can tell you would quickly run out of available I/O pins.

There are two solutions to this problem. One solution is to use a 7447 or similar BCD to seven segment decoder driver. The other solution is to use "multiplexing". With multiplexing you first light up one seven segment display and then the other one. This multiplexing is done so rapidly that they both appear to be lit all the time.

Later on we will be using some "multiplexing" to switch between multiple seven segment displays. That way only nine Arduino pins can control two seven segment LED arrays so we can display all the numbers from zero to 99. Without the multiplexing that would take all 14 digital I/O pins! Multiplexing will be used a lot more later on when we start using LED arrays.

Here is how the segments are labeled and the pins are connected (as seen from above) for a FND503 seven segment common cathode display.

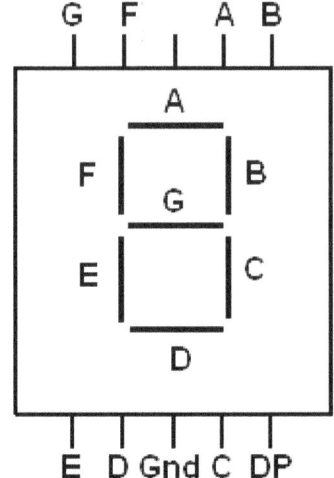

To display the numbers we need to turn on the correct segments like this.

0 – A B C D E F
1 – B C
2 – A B G E D
3 – A B C D G
4 – B C F G
5 – A C D F G
6 – A C D E F G
7 – A B C
8 – A B C D E F G
9 – A B C F G

Here is a list of the needed parts to make this work, besides the Arduino itself.

1. One Common Cathode, Seven Segment LED display.
2. Seven 470 ohm resistors.

Here is the schematic diagram for a one digit counter, note that the wires to F and G must be crossed:

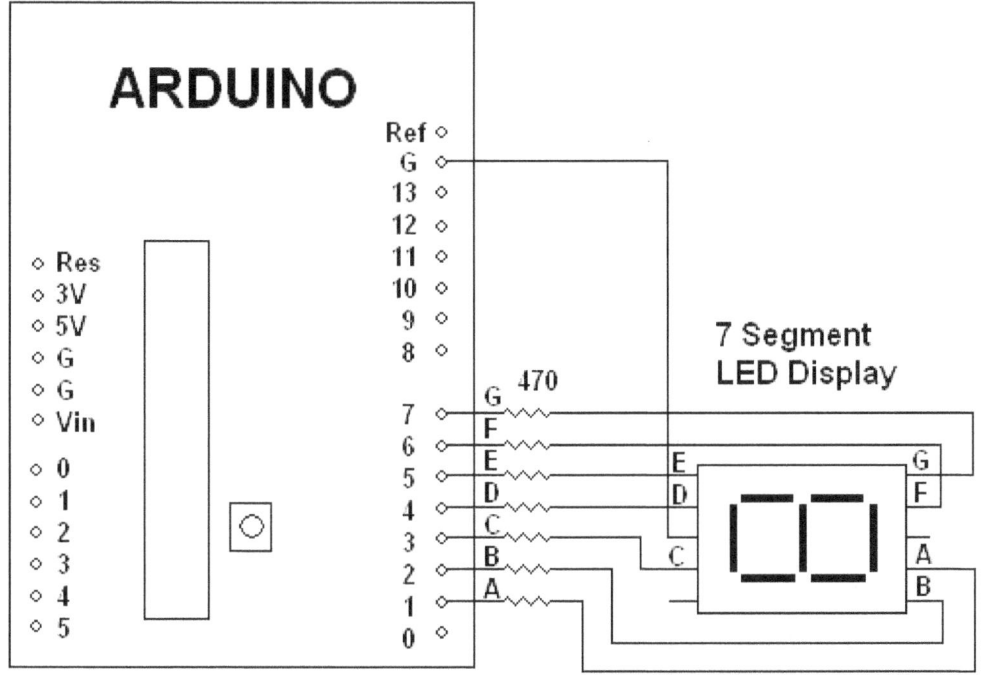

Here is the software to make a one digit counter:

```
//*****************************************/
//  Name    : LED 7 Segment Counter
//  Author  : Bob Davis
//  Date    : 12 March, 2013
//  Version : 1.0
//*****************************************/
// Pins for the Seven Segments
int SegA = 1;
int SegB = 2;
int SegC = 3;
int SegD = 4;
int SegE = 5;
int SegF = 6;
int SegG = 7;
int DispA = 8;
int DispB = 9;

// Set the pins to output to the LED's
```

```
void setup() {
  pinMode(SegA, OUTPUT);
  pinMode(SegB, OUTPUT);
  pinMode(SegC, OUTPUT);
  pinMode(SegD, OUTPUT);
  pinMode(SegE, OUTPUT);
  pinMode(SegF, OUTPUT);
  pinMode(SegG, OUTPUT);
  pinMode(DispA, OUTPUT);
  pinMode(DispB, OUTPUT);
}
void loop() {
  for (int count = 0; count < 10; count++){
    digitalWrite (SegA, LOW);
    digitalWrite (SegB, LOW);
    digitalWrite (SegC, LOW);
    digitalWrite (SegD, LOW);
    digitalWrite (SegE, LOW);
    digitalWrite (SegF, LOW);
    digitalWrite (SegG, LOW);
    if (count == 1){
      digitalWrite (SegB, HIGH);
      digitalWrite (SegC, HIGH);
      }
    if (count == 2){
      digitalWrite (SegA, HIGH);
      digitalWrite (SegB, HIGH);
      digitalWrite (SegD, HIGH);
      digitalWrite (SegE, HIGH);
      digitalWrite (SegG, HIGH);
      }
    if (count == 3){
      digitalWrite (SegA, HIGH);
      digitalWrite (SegB, HIGH);
      digitalWrite (SegC, HIGH);
      digitalWrite (SegD, HIGH);
      digitalWrite (SegG, HIGH);
      }
    if (count == 4){
      digitalWrite (SegB, HIGH);
      digitalWrite (SegC, HIGH);
```

```
    digitalWrite (SegF, HIGH);
    digitalWrite (SegG, HIGH);
    }
if (count == 5){
    digitalWrite (SegA, HIGH);
    digitalWrite (SegC, HIGH);
    digitalWrite (SegD, HIGH);
    digitalWrite (SegF, HIGH);
    digitalWrite (SegG, HIGH);
    }
if (count == 6){
    digitalWrite (SegA, HIGH);
    digitalWrite (SegC, HIGH);
    digitalWrite (SegD, HIGH);
    digitalWrite (SegE, HIGH);
    digitalWrite (SegF, HIGH);
    digitalWrite (SegG, HIGH);
    }
if (count == 7){
    digitalWrite (SegA, HIGH);
    digitalWrite (SegB, HIGH);
    digitalWrite (SegC, HIGH);
    }
if (count == 8){
    digitalWrite (SegA, HIGH);
    digitalWrite (SegB, HIGH);
    digitalWrite (SegC, HIGH);
    digitalWrite (SegD, HIGH);
    digitalWrite (SegE, HIGH);
    digitalWrite (SegF, HIGH);
    digitalWrite (SegG, HIGH);
    }
if (count == 9){
    digitalWrite (SegA, HIGH);
    digitalWrite (SegB, HIGH);
    digitalWrite (SegC, HIGH);
    digitalWrite (SegD, HIGH);
    digitalWrite (SegF, HIGH);
    digitalWrite (SegG, HIGH);
    }
if (count == 0){
```

```
    digitalWrite (SegA, HIGH);
    digitalWrite (SegB, HIGH);
    digitalWrite (SegC, HIGH);
    digitalWrite (SegD, HIGH);
    digitalWrite (SegE, HIGH);
    digitalWrite (SegF, HIGH);
    }

  delay(1000);
}}
```

Chapter 11

Meter/Temperature Display

We can use one of the analog inputs combined with using two seven segment displays to make a simple volt meter. We can then use that meter to measure light, temperature, sound levels, and just about anything else that can be measured as a voltage. For this project we will add a LM34 temperature sensor and display the room temperature on the two seven segment LED displays.

It is possible to even build you own energy management system using the basics that are shown in this chapter. Combining multiple temperature sensors with the dual seven segment drivers you can now display temperatures ranging from zero to 99 degrees. If you were to add relays you could then turn the heat on and off based on the room temperature.

Here is a list of the needed parts to make this work, besides the Arduino itself.

1. Two Common Cathode, Seven Segment LED's.
2. Seven 470 ohm resistors.
3. One LM34 temperature sensor.

Up next is the schematic wiring diagram for the two digit meter with a multiplexed dispaly.

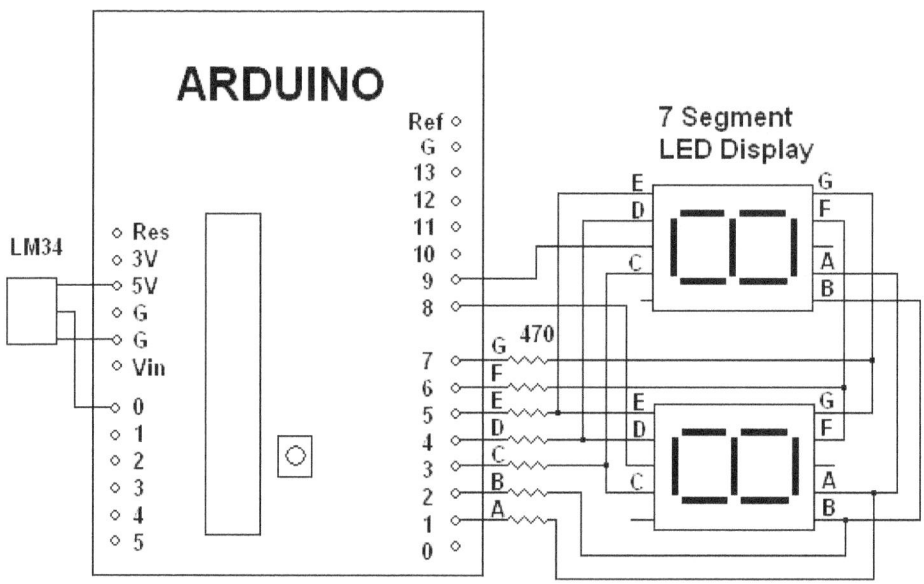

Here is a picture of what it looks like when it is all wired up and working. Note the lm34 over on the left side of the breadboard.

Here is the LM34 connection diagram as viewed from below. The LM34 outputs the temperature at .01 volts per degree Fahrenheit. That is to say that 70 degrees would be seen as .70 volts out of the LM35.

LM-34
TO-92 Package

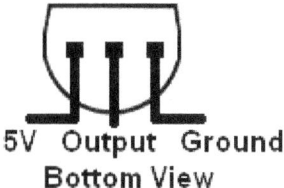

5V Output Ground
Bottom View

Here is the sketch or code to make it work;

```
//***************************************/
// Name    : LED Meter Driver
// Author  : Bob Davis
// Date    : 12 March, 2013
// Version : 1.0
//***************************************/
// Pins for the Seven Segments
int SegA = 1;
int SegB = 2;
int SegC = 3;
int SegD = 4;
int SegE = 5;
int SegF = 6;
int SegG = 7;
int DispA = 8;
int DispB = 9;
byte temp = 0;
byte dig1; // left digit
byte dig2; // right digit
int pass = 0;
int digit;
// Set the pins to output to the LED's
void setup() {
  pinMode(SegA, OUTPUT);
  pinMode(SegB, OUTPUT);
  pinMode(SegC, OUTPUT);
  pinMode(SegD, OUTPUT);
```

```
  pinMode(SegE, OUTPUT);
  pinMode(SegF, OUTPUT);
  pinMode(SegG, OUTPUT);
  pinMode(DispA, OUTPUT);
  pinMode(DispB, OUTPUT);
}
void loop() {
  // 5V=1024 so if you divide by 2 then 5V=512
  // 70 degrees is .7 volts is 71.62
  temp = analogRead(0)/2;
  dig1 = temp / 10;  // gives firs digit
  dig2 = (temp % 10); // gives remainder
  if (pass == 0){
    digitalWrite (DispA, LOW);
    digitalWrite (DispB, HIGH);
    digit = dig1;
    pass = 1;
  }
  else {
    digitalWrite (DispA, HIGH);
    digitalWrite (DispB, LOW);
    digit = dig2;
    pass = 0;
  }
    digitalWrite (SegA, LOW);
    digitalWrite (SegB, LOW);
    digitalWrite (SegC, LOW);
    digitalWrite (SegD, LOW);
    digitalWrite (SegE, LOW);
    digitalWrite (SegF, LOW);
    digitalWrite (SegG, LOW);
    if (digit == 1){
      digitalWrite (SegB, HIGH);
      digitalWrite (SegC, HIGH);
      }
    if (digit == 2){
      digitalWrite (SegA, HIGH);
      digitalWrite (SegB, HIGH);
      digitalWrite (SegD, HIGH);
      digitalWrite (SegE, HIGH);
      digitalWrite (SegG, HIGH);
```

```
  }
if (digit == 3){
  digitalWrite (SegA, HIGH);
  digitalWrite (SegB, HIGH);
  digitalWrite (SegC, HIGH);
  digitalWrite (SegD, HIGH);
  digitalWrite (SegG, HIGH);
  }
if (digit == 4){
  digitalWrite (SegB, HIGH);
  digitalWrite (SegC, HIGH);
  digitalWrite (SegF, HIGH);
  digitalWrite (SegG, HIGH);
  }
if (digit == 5){
  digitalWrite (SegA, HIGH);
  digitalWrite (SegC, HIGH);
  digitalWrite (SegD, HIGH);
  digitalWrite (SegF, HIGH);
  digitalWrite (SegG, HIGH);
  }
if (digit == 6){
  digitalWrite (SegA, HIGH);
  digitalWrite (SegC, HIGH);
  digitalWrite (SegD, HIGH);
  digitalWrite (SegE, HIGH);
  digitalWrite (SegF, HIGH);
  digitalWrite (SegG, HIGH);
  }
if (digit == 7){
  digitalWrite (SegA, HIGH);
  digitalWrite (SegB, HIGH);
  digitalWrite (SegC, HIGH);
  }
if (digit == 8){
  digitalWrite (SegA, HIGH);
  digitalWrite (SegB, HIGH);
  digitalWrite (SegC, HIGH);
  digitalWrite (SegD, HIGH);
  digitalWrite (SegE, HIGH);
  digitalWrite (SegF, HIGH);
```

```
      digitalWrite (SegG, HIGH);
    }
  if (digit == 9){
    digitalWrite (SegA, HIGH);
    digitalWrite (SegB, HIGH);
    digitalWrite (SegC, HIGH);
    digitalWrite (SegD, HIGH);
    digitalWrite (SegF, HIGH);
    digitalWrite (SegG, HIGH);
    }
  if (digit == 0){
    digitalWrite (SegA, HIGH);
    digitalWrite (SegB, HIGH);
    digitalWrite (SegC, HIGH);
    digitalWrite (SegD, HIGH);
    digitalWrite (SegE, HIGH);
    digitalWrite (SegF, HIGH);
    }
  delay(10);
}
```

Chapter 12

Three Digit Meter

What happens when the temperature in the last project exceeds 99 degrees? Adding a third digit would solve that problem. For this project we will just add another seven segment LED display to the last project. The math gets a little bit tricky to get the three digits separated out to display them one at a time. The picture below shows a LM34 connected to the meter.

We will also add a UGN3503 magnetic field sensor. Be careful because the ground and output pins are different than the LM34 temperature sensor. Below is a schematic of the pin connections for the UGN3503

UGN3503
Hall Effect

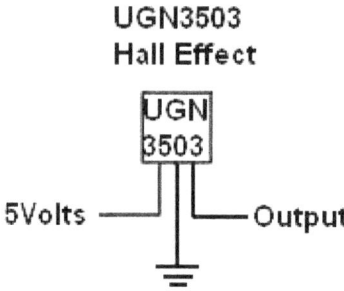

Here is the schematic diagram of the Arduino powered three digit meter.

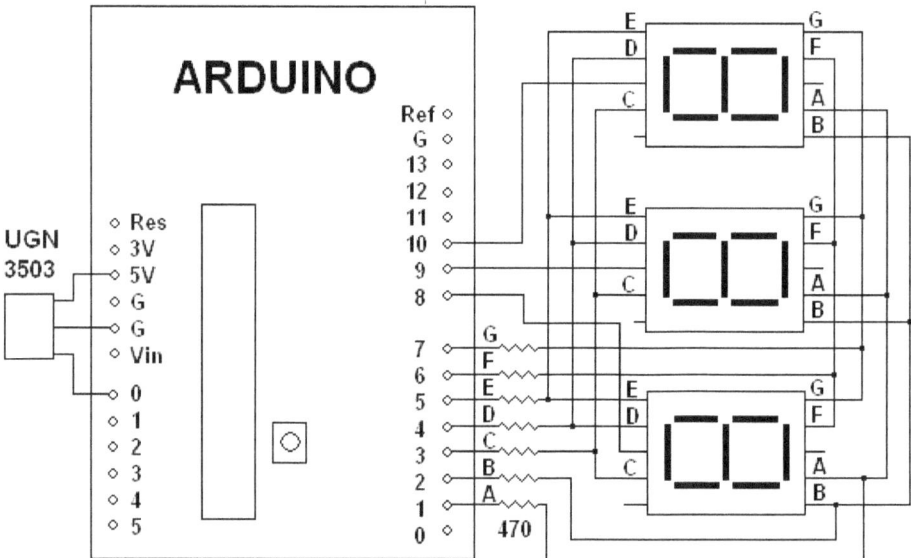

Here is a list of the needed parts to make this work, besides the Arduino itself.

4. Three Common Cathode, Seven Segment LED's.
5. Seven 470 ohm resistors.
6. One UGN3503 magnetic field sensor.

Here is the code to make the three digit meter work.

```
//********************************************//
//  Name    : LED Meter Driver 3       //
//  Author  : Bob Davis                //
```

```
//  Date    : 12 March, 2013                    //
//  Version : 1.0                               //
//*******************************************//
// Pins for the Seven Segments
int SegA = 1;
int SegB = 2;
int SegC = 3;
int SegD = 4;
int SegE = 5;
int SegF = 6;
int SegG = 7;
int DispA = 8;
int DispB = 9;
int DispC = 10;
int temp = 0;
byte dig1;
byte dig2;
byte dig3;
int pass = 0;
int digit;
// Set the pins to output to the LED's
void setup() {
  pinMode(SegA, OUTPUT);
  pinMode(SegB, OUTPUT);
  pinMode(SegC, OUTPUT);
  pinMode(SegD, OUTPUT);
  pinMode(SegE, OUTPUT);
  pinMode(SegF, OUTPUT);
  pinMode(SegG, OUTPUT);
  pinMode(DispA, OUTPUT);
  pinMode(DispB, OUTPUT);
  pinMode(DispC, OUTPUT);
}
void loop() {
  // 5V=1024 so if you divide by 2.05 then 5V=500
  // 70 degrees is .7 volts is 070
  temp = analogRead(0)/2.05;
  dig1 = (temp % 10);   // gives right digit
  dig2 = (temp / 10) % 10;  // gives middle digit
  dig3 = temp / 100;    // gives left digit
  pass++;
```

```
if (pass == 1){
  digitalWrite (DispA, LOW);
  digitalWrite (DispB, HIGH);
  digitalWrite (DispC, HIGH);
  digit = dig1;
    }
if (pass == 2) {
  digitalWrite (DispA, HIGH);
  digitalWrite (DispB, LOW);
  digitalWrite (DispC, HIGH);
  digit = dig2;
    }
if (pass == 3) {
  digitalWrite (DispA, HIGH);
  digitalWrite (DispB, HIGH);
  digitalWrite (DispC, LOW);
  digit = dig3;
  pass = 0;
    }
  digitalWrite (SegA, LOW);
  digitalWrite (SegB, LOW);
  digitalWrite (SegC, LOW);
  digitalWrite (SegD, LOW);
  digitalWrite (SegE, LOW);
  digitalWrite (SegF, LOW);
  digitalWrite (SegG, LOW);
  if (digit == 1){
    digitalWrite (SegB, HIGH);
    digitalWrite (SegC, HIGH);
    }
  if (digit == 2){
    digitalWrite (SegA, HIGH);
    digitalWrite (SegB, HIGH);
    digitalWrite (SegD, HIGH);
    digitalWrite (SegE, HIGH);
    digitalWrite (SegG, HIGH);
    }
  if (digit == 3){
    digitalWrite (SegA, HIGH);
    digitalWrite (SegB, HIGH);
```

```
  digitalWrite (SegC, HIGH);
  digitalWrite (SegD, HIGH);
  digitalWrite (SegG, HIGH);
  }
if (digit == 4){
  digitalWrite (SegB, HIGH);
  digitalWrite (SegC, HIGH);
  digitalWrite (SegF, HIGH);
  digitalWrite (SegG, HIGH);
  }
if (digit == 5){
  digitalWrite (SegA, HIGH);
  digitalWrite (SegC, HIGH);
  digitalWrite (SegD, HIGH);
  digitalWrite (SegF, HIGH);
  digitalWrite (SegG, HIGH);
  }
if (digit == 6){
  digitalWrite (SegA, HIGH);
  digitalWrite (SegC, HIGH);
  digitalWrite (SegD, HIGH);
  digitalWrite (SegE, HIGH);
  digitalWrite (SegF, HIGH);
  digitalWrite (SegG, HIGH);
  }
if (digit == 7){
  digitalWrite (SegA, HIGH);
  digitalWrite (SegB, HIGH);
  digitalWrite (SegC, HIGH);
  }
if (digit == 8){
  digitalWrite (SegA, HIGH);
  digitalWrite (SegB, HIGH);
  digitalWrite (SegC, HIGH);
  digitalWrite (SegD, HIGH);
  digitalWrite (SegE, HIGH);
  digitalWrite (SegF, HIGH);
  digitalWrite (SegG, HIGH);
  }
if (digit == 9){
  digitalWrite (SegA, HIGH);
```

```
  digitalWrite (SegB, HIGH);
  digitalWrite (SegC, HIGH);
  digitalWrite (SegD, HIGH);
  digitalWrite (SegF, HIGH);
  digitalWrite (SegG, HIGH);
  }
 if (digit == 0){
  digitalWrite (SegA, HIGH);
  digitalWrite (SegB, HIGH);
  digitalWrite (SegC, HIGH);
  digitalWrite (SegD, HIGH);
  digitalWrite (SegE, HIGH);
  digitalWrite (SegF, HIGH);
  }
 if (digit >= 10){   //out of range
  digitalWrite (SegG, HIGH);
  }
 delay(5);
}
```

Chapter 13

Eight by Eight LED Array

An eight by eight LED display would require 16 pins to connect it to the Arduino. There are several possible solutions. You can use any or a combination of the following solutions to save on the Arduino pins. One solution is to use the analog input pins as output pins. That can be a problem if you want to display some data from the analog inputs on the LED array. We will do that for this project.

To Control the LED Array's rows you can use a 74HC138 three to eight de-multiplexer to allow three pins from the Arduino to select between the eight rows. The 74HC138 has active low outputs and you usually want active high outputs. That problem can be fixed by using PNP driver transistors or by using an IC that has eight inverters. Another solution is to use a 4017 counter and decoder. The 4017 decodes to 10 outputs, with just a clock input and has active high outputs. However the 4017 has almost no drive power so a driver IC or transistors will be needed.

To control the columns you can use a 74HC595 or similar shift registers. This will allow two pins to send clock and data to the shift register that will then be connected to the eight columns. A shift register is a series of eight latches; every time it gets a "clock" it shifts the data down one position to the next latch.

For this first project we will not be using any support chips. That essentially limits us to an eight by eight LED display. It is possible to use the analog input pins as outputs. You can add support chips to get more than an eight by eight display working, and we will do that in the next project.

Using LED matrix arrays used to require the use of something that was called a "Character Generator". The characters we will use will be created in software using "binary" values. Basically you select what LED's you want to light up. Then you draw them on a chart. Next you add up their binary values you will get the binary value for each row of the character.

Below is a chart showing how to create the letter "A", a "@" is a lit LED and a "O" is a LED that is not lit.

16	8	4	2	1	Binary Values
O	O	@	O	O	4
O	@	O	@	O	10
@	O	O	O	@	17
@	@	@	@	@	31
@	O	O	O	@	17
@	O	O	O	@	17
@	O	O	O	@	17

This would show up like this in the sketch code:
bitmap [8]={0, 4, 10, 17, 31, 17, 17, 17};

Since we are working with eight by eight arrays, lets use some more LED's. Below is a chart showing how to create a smiley face, again a "@" is a lit LED.

128	64	32	16	8	4	2	1	Binary Value
O	O	@	@	@	@	O	O	60
O	@	O	O	O	O	@	O	66
@	O	@	O	O	@	O	@	165
@	O	O	O	O	O	O	@	129
@	O	@	O	O	@	O	@	165
@	O	O	@	@	O	O	@	153
O	@	O	O	O	O	@	O	66
O	O	@	@	@	@	O	O	60

This would show up like this in the sketch code:
bitmap [8]={60, 66, 165, 129, 165, 153, 66, 60};

Up next is a schematic showing the pin connections for a LTP14188A eight by eight dual color common anode LED array that we will be using for the next few projects. It is easy to wire up because the pins are in a nice sequential order instead of alternating like most other LED arrays. The "Column" pins are active when high, the "Rows" light up when low.

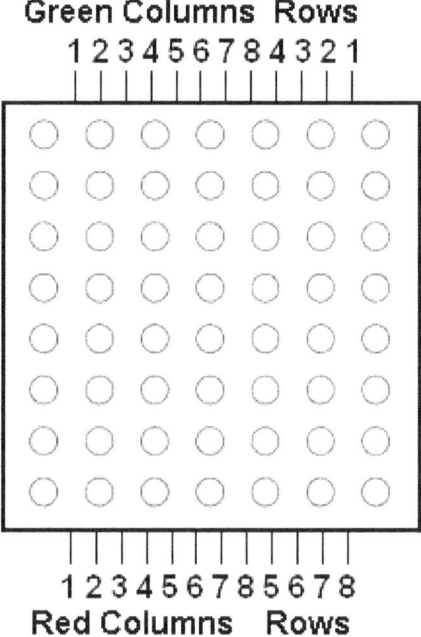

Here is a list of the needed parts to make this work, besides the Arduino itself.

1. One five by eight or eight by eight LED array,
   The LTP14188 is recommended because it is easy to wire up.
2. Eight 100 ohm resistors.

Note that we have changed to using 100 ohm resistors. Now that we are "multiplexing" each LED is only on for 1/8 of the time. So the resistor must be smaller so that more current will flow.

Here is the schematic diagram for wiring up an eight by eight green display without using any support IC's. Note the use of the analog inputs as outputs. Also there are still four unused pins left over that can be used for some sort of inputs if needed.

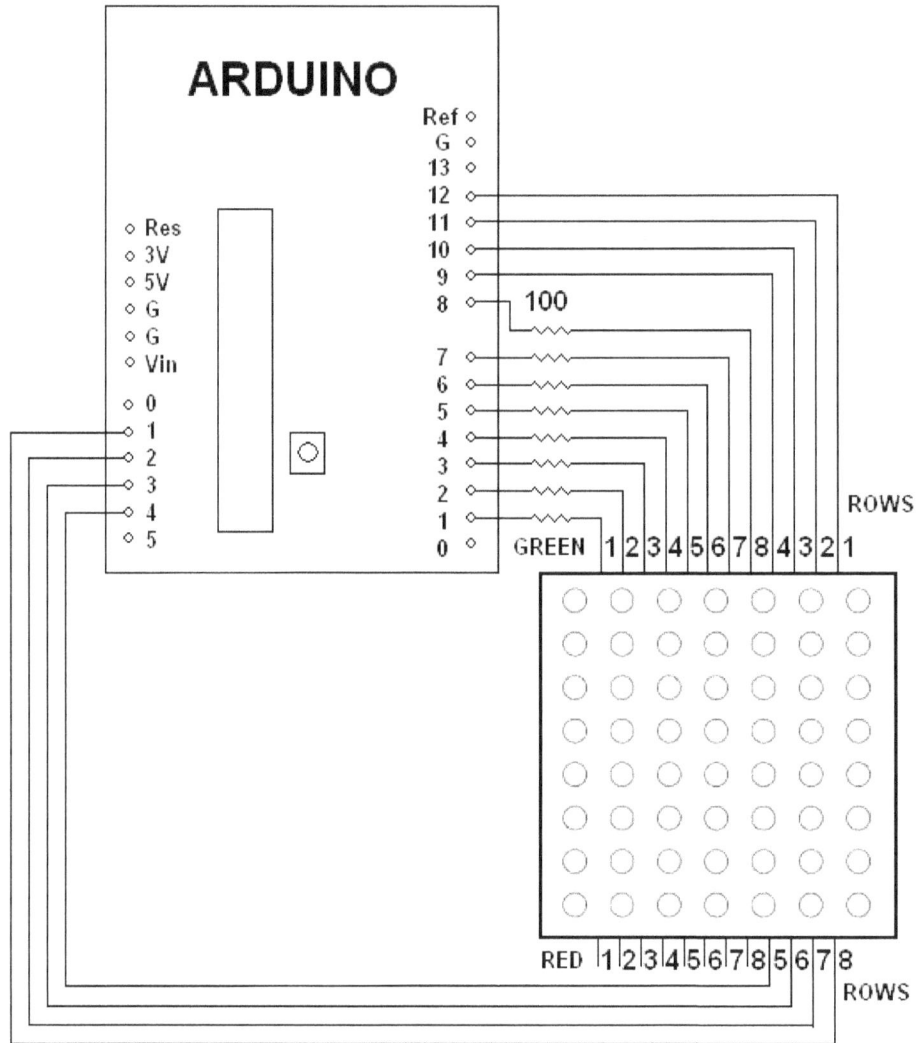

Here is a picture of what it could look like when it is all wired up. Note that the wires coming from the side of the display that is closest to you are actually coming from underneath the LED array. The spacing for the breadboard makes things a little too tight to easily get to both sides of the LED array. Also I was using 220 ohm resistors and I would recommend that you use 47 or 100 ohm resistors for more brightness.

Here is the sketch code to display the smiley face;

```
//*****************************************/
// Name    : LED 8x8 array Driver
// Author  : Bob Davis
// Date    : 12 March, 2013
// Version : 1.0
//*****************************************/
// Pins for the LED Array
int Col1 = 1;
int Col2 = 2;
int Col3 = 3;
int Col4 = 4;
int Col5 = 5;
int Col6 = 6;
int Col7 = 7;
int Col8 = 8;
int Row4 = 9;
```

```
int Row3 = 10;
int Row2 = 11;
int Row1 = 12;
int Row8 = A1;
int Row7 = A2;
int Row6 = A3;
int Row5 = A4;
// array to make a smiley face
byte bitmap [8]={60, 66, 165, 129, 165, 153, 66, 60};

// Set the pins to output to the array
void setup() {
  pinMode(Col1, OUTPUT);
  pinMode(Col2, OUTPUT);
  pinMode(Col3, OUTPUT);
  pinMode(Col4, OUTPUT);
  pinMode(Col5, OUTPUT);
  pinMode(Col6, OUTPUT);
  pinMode(Col7, OUTPUT);
  pinMode(Col8, OUTPUT);
  pinMode(Row1, OUTPUT);
  pinMode(Row2, OUTPUT);
  pinMode(Row3, OUTPUT);
  pinMode(Row4, OUTPUT);
  pinMode(Row5, OUTPUT);
  pinMode(Row6, OUTPUT);
  pinMode(Row7, OUTPUT);
  pinMode(Row8, OUTPUT);
}
void loop() {
  for (int row = 0; row <8; row++){
    //turn everything off
    digitalWrite (Col1, HIGH);
    digitalWrite (Col2, HIGH);
    digitalWrite (Col3, HIGH);
    digitalWrite (Col4, HIGH);
    digitalWrite (Col5, HIGH);
    digitalWrite (Col6, HIGH);
    digitalWrite (Col7, HIGH);
    digitalWrite (Col8, HIGH);
    digitalWrite (Row1, LOW);
```

```
    digitalWrite (Row2, LOW);
    digitalWrite (Row3, LOW);
    digitalWrite (Row4, LOW);
    digitalWrite (Row5, LOW);
    digitalWrite (Row6, LOW);
    digitalWrite (Row7, LOW);
    digitalWrite (Row8, LOW);
    // set up the data in the columns
    if bitRead(bitmap[row],0) digitalWrite (Col1, LOW);
    if bitRead(bitmap[row],1) digitalWrite (Col2, LOW);
    if bitRead(bitmap[row],2) digitalWrite (Col3, LOW);
    if bitRead(bitmap[row],3) digitalWrite (Col4, LOW);
    if bitRead(bitmap[row],4) digitalWrite (Col5, LOW);
    if bitRead(bitmap[row],5) digitalWrite (Col6, LOW);
    if bitRead(bitmap[row],6) digitalWrite (Col7, LOW);
    if bitRead(bitmap[row],7) digitalWrite (Col8, LOW);
    // turn on the associated row
    if (row == 0) digitalWrite (Row1, HIGH);
    if (row == 1) digitalWrite (Row2, HIGH);
    if (row == 2) digitalWrite (Row3, HIGH);
    if (row == 3) digitalWrite (Row4, HIGH);
    if (row == 4) digitalWrite (Row5, HIGH);
    if (row == 5) digitalWrite (Row6, HIGH);
    if (row == 6) digitalWrite (Row7, HIGH);
    if (row == 7) digitalWrite (Row8, HIGH);
    delay(3);
} }
```

Chapter 14

Dual Color Eight by Eight

LED Array

If we were to add some shift registers to the previous design then we could control a second set of eight columns. This addition allows for a second color to be displayed besides the usual red color. Usually the second color is green, but we started with that color first. With the addition of the two shift registers we free up a lot of the Arduino pins, while adding more features to the display.

Here is a list of the needed parts to make this work, besides the Arduino itself.

1. One dual color eight by eight LED array.
2. Sixteen 100 or 220 ohm resistors.
3. Two 74HC595 shift register.

Since we are working with dual color arrays, we need two bitmaps, one for each color. Below is a chart showing how to create the outside of the face in green, a "@" is a lit LED.

128	64	32	16	8	4	2	1	Binary Value
O	O	@	@	@	@	O	O	60
O	@	O	O	O	O	@	O	66
@	O	O	O	O	O	O	@	129
@	O	O	O	O	O	O	@	129
@	O	O	O	O	O	O	@	129
@	O	O	O	O	O	O	@	129
O	@	O	O	O	O	@	O	66
O	O	@	@	@	@	O	O	60

This array would show up as this in the sketch code:
bitmap [8]={60, 66, 129, 129, 129, 129, 66, 60};

Below is a chart showing how to create the mouth and eyes in red, a "@"
is a lit LED.

128	64	32	16	8	4	2	1	Binary Value
O	O	@	@	@	@	O	O	60
O	@	O	O	O	O	@	O	66
O	O	@	O	O	@	O	O	36
O	O	O	O	O	O	O	O	0
O	O	@	O	O	@	O	O	36
O	O	O	@	@	O	O	O	24
O	O	O	O	O	O	O	O	0
O	O	O	O	O	O	O	O	0

This would show up like this in the sketch code:
bitmap [8]={60, 66, 36, 0, 36, 24, 0, 0};

Coming up next is the schematic for a dual color version using 74HC595
shift registers.

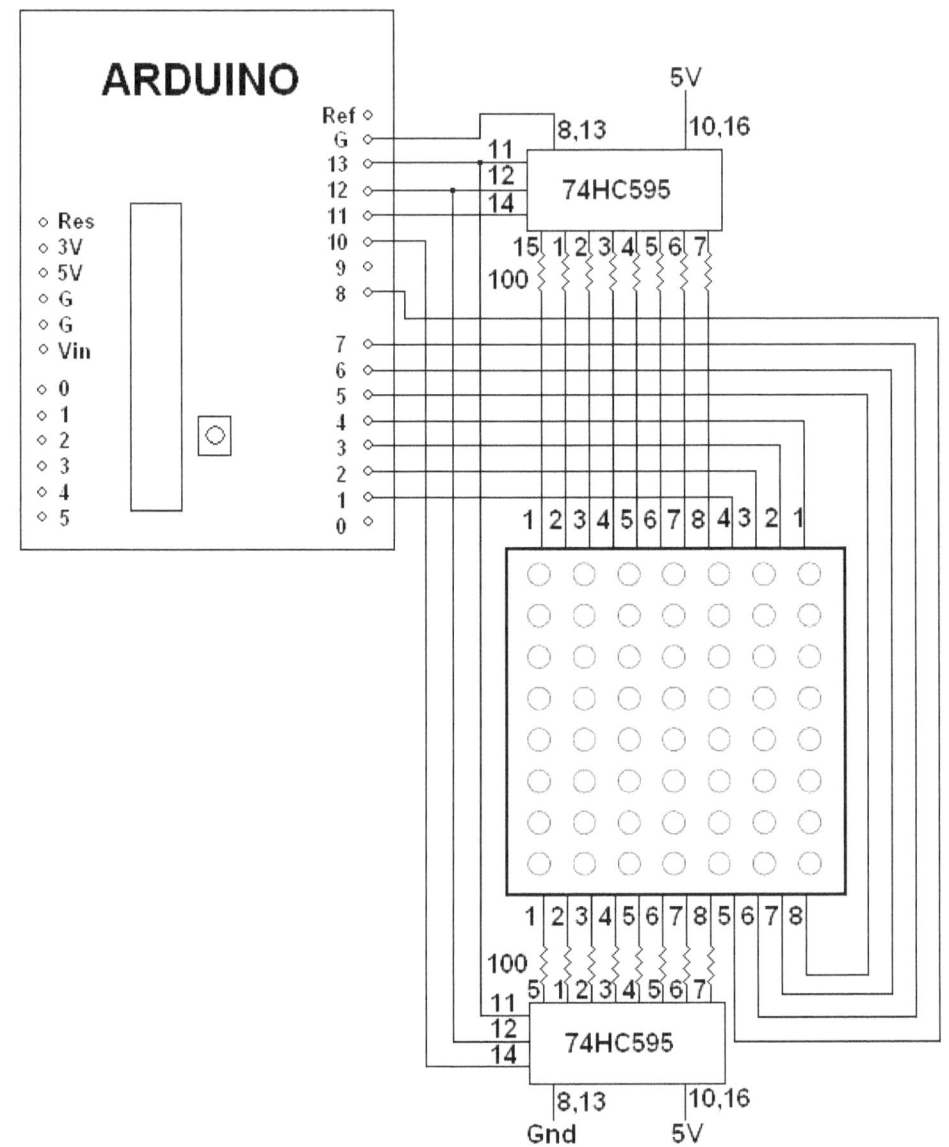

Here is the top view of the wiring diagram for the 74HC595. Below that is what it usually looks like in a schematic. First observe that pin one is found to the left of a notch in the chip. Note that the orientation of the pins does not usually match the schematic. It is hard to show IC's in a schematic exactly like they are physically wired up.

Here is the code for the sketch to make it work:

```
//*******************************************/
// Name    : LED 8x8 SR array Driver
// Author  : Bob Davis
// Date    : 12 March, 2013
// Version : 1.0
//*******************************************/
// Pins for the Columns and Rows
int Row1 = 4;
int Row2 = 3;
int Row3 = 2;
int Row4 = 1;
int Row5 = 5;
int Row6 = 6;
int Row7 = 7;
int Row8 = 8;
int GSerial = 10;
int RSerial = 11;
int RCLK = 12;
int SCLK = 13;
// arrays to make a red and green smiley face
```

```
byte Gbitmap [8]={60, 66, 129, 129, 129, 129, 66, 60};
byte Rbitmap [8]={60, 66, 36, 0, 36, 24, 0, 0};
// Set the pins to output to the array
void setup() {
  pinMode(Row1, OUTPUT);
  pinMode(Row2, OUTPUT);
  pinMode(Row3, OUTPUT);
  pinMode(Row4, OUTPUT);
  pinMode(Row5, OUTPUT);
  pinMode(Row6, OUTPUT);
  pinMode(Row7, OUTPUT);
  pinMode(Row8, OUTPUT);
  pinMode(GSerial, OUTPUT);
  pinMode(RSerial, OUTPUT);
  pinMode(RCLK, OUTPUT);
  pinMode(SCLK, OUTPUT);
}
void loop() {
  for (int row = 0; row <8; row++){
    //turn everything off
    digitalWrite (Row1, LOW);
    digitalWrite (Row2, LOW);
    digitalWrite (Row3, LOW);
    digitalWrite (Row4, LOW);
    digitalWrite (Row5, LOW);
    digitalWrite (Row6, LOW);
    digitalWrite (Row7, LOW);
    digitalWrite (Row8, LOW);
    // set up the data in the columns
    for (int shiftbit = 8; shiftbit >=0; shiftbit--){
    digitalWrite(GSerial, HIGH);
    digitalWrite(RSerial, HIGH);
    if bitRead(Gbitmap[row],shiftbit) digitalWrite(GSerial, LOW);
    if bitRead(Rbitmap[row],shiftbit) digitalWrite(RSerial, LOW);
    digitalWrite(SCLK, HIGH);
    digitalWrite(SCLK, LOW);
      }
    //latch the data
    digitalWrite(RCLK, HIGH);
    digitalWrite(RCLK, LOW);
    // turn on the associated row
```

```
    if (row == 0) digitalWrite (Row1, HIGH);
    if (row == 1) digitalWrite (Row2, HIGH);
    if (row == 2) digitalWrite (Row3, HIGH);
    if (row == 3) digitalWrite (Row4, HIGH);
    if (row == 4) digitalWrite (Row5, HIGH);
    if (row == 5) digitalWrite (Row6, HIGH);
    if (row == 6) digitalWrite (Row7, HIGH);
    if (row == 7) digitalWrite (Row8, HIGH);
    delay(2);
}  }
```

Chapter 15

Dual Color Eight by 15 LED Array

This project will make an Arduino powered eight by 15 red and green LED sign. You can also get orange by turning on red and green at the same time. You can start with one array for eight by five and then expand it as you get each section working. This project demonstrates chaining shift registers together by connecting serial out, pin nine, to serial in, pin 14, of the next shift register.

This time we will be using several 1702054 eight by five LED arrays. It takes three of them to make the 8 by 15 sign. It is possible to take this design on to make a much larger LED sign. Once again I set down with a nine volt battery and a 1K resistor and figured out the pin connections for this display. It is actually a very common arrangement.

Here is the pin out for one of the LED arrays. The rows are called "Dx" so they will not be confused with the red columns:

```
R1  D3  D1  D2  D4  G3  G4  R5  G5
---------------------------------------------
      O   O   O   O   O
      O   O   O   O   O
      O   O   O   O   O
      O   O   O   O   O
      O   O   O   O   O
      O   O   O   O   O
      O   O   O   O   O
      O   O   O   O   O
---------------------------------------------
G1  G2  R2  R3  D6  D7  R4  D5  D8
```

Now to get an eight by 15 display you need to use three of these LED arrays, so their pin definitions will have to be rearranged. Here are the pin definitions for the next two LED arrays:

R6 D3 D1 D2 D4 G8 G9 R10 G10

G6 G7 R7 R8 D6 D7 R9 D5 D8

R11 D3 D1 D2 D4 G13 G14 R15 G15

G11 G12 R12 R13 D6 D7 R14 D5 D8

This is a schematic diagram that shows all three of the eight by five LED arrays. It had to be shrunk to fit so the letters might be a little hard to read.

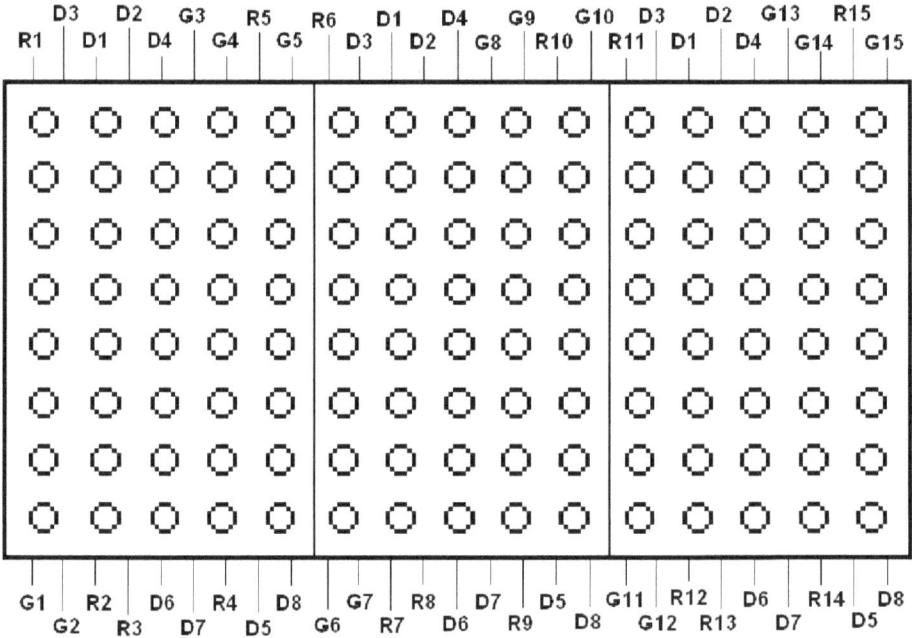

Up next is a schematic diagram of the eight by 15 LED display circuitry. It is too difficult to show the LED array pins at their actual physical locations. That is why there is a separate schematic of the LED arrays. The Arduino had to be cut off on the left to fit this schematic at its actual size.

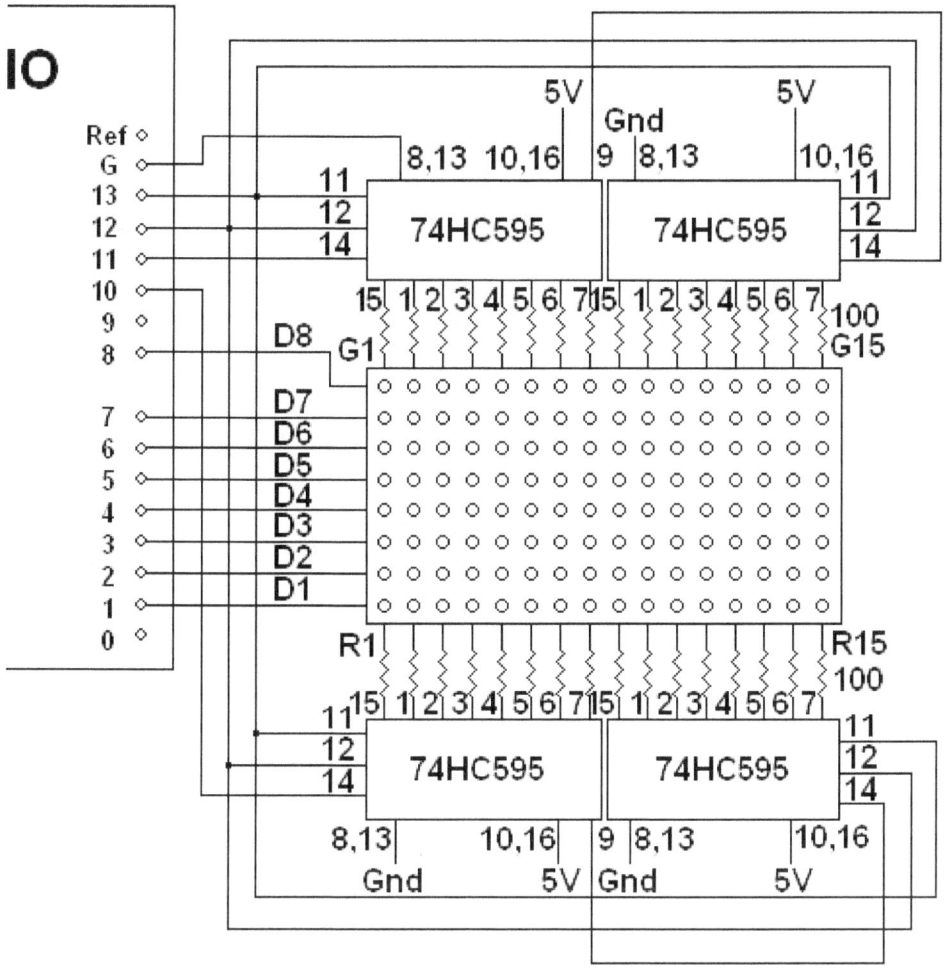

Here is what the display looks like when it is assembled and working. Note that the parts just so fit on standard seven inch long breadboards. The biggest problem I had was getting the resistors to align with the pins of the LED arrays. Twice I was off by one position.

Also the LED's are not very bright because there are no row drivers and the Arduino has a hard time sourcing the necessary amount of current. Eight NPN driver transistors such as TIP120 or TIP121's can be used on the rows to make the display a little brighter. They are visible on a smaller breadboard below the Arduino. I soldered their collectors together and connected them to five volts. On the left, their base goes to the Arduino. On the right, their emitter goes to the LED arrays.

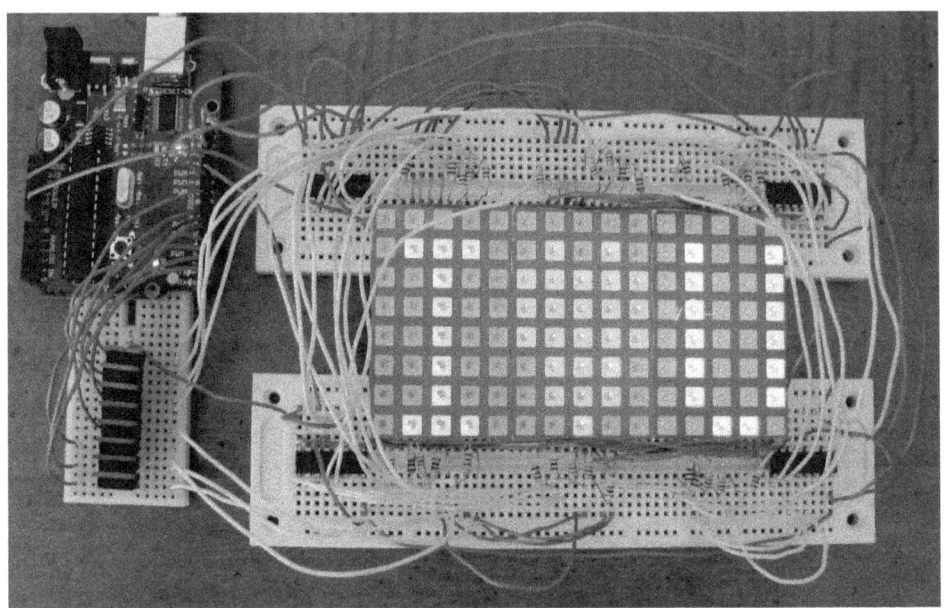

The serial data is converted to parallel by four 74HC595 shift registers. One is located at each end of each breadboard. It is possible to have the resistors go directly from the shift registers to the arrays, but it is less confusing if jumper wires are used. This project used all of my jumper wires and I had to make a few more jumpers to complete it.

Here is a sketch to display 3 letters as seen in the picture above.

```
//****************************************//
// Name    : Breadboard 3 digit dual color        //
// Author  : Bob Davis                    //
// Date    : 12 January, 2014              //
// Version : 1.0                           //
//****************************************//
// Pins for the row drivers
int rowPin1 = 1;
int rowPin2 = 2;
int rowPin3 = 3;
int rowPin4 = 4;
int rowPin5 = 5;
int rowPin6 = 6;
int rowPin7 = 7;
int rowPin8 = 8;
int gdataPin = 10;
```

```
int rdataPin = 11;
int rclockPin = 12;
int clockPin = 13;
// Set the pins to output to the sign
void setup() {
  pinMode(rowPin1, OUTPUT);
  pinMode(rowPin2, OUTPUT);
  pinMode(rowPin3, OUTPUT);
  pinMode(rowPin4, OUTPUT);
  pinMode(rowPin5, OUTPUT);
  pinMode(rowPin6, OUTPUT);
  pinMode(rowPin7, OUTPUT);
  pinMode(rowPin8, OUTPUT);
  pinMode(rclockPin, OUTPUT);
  pinMode(clockPin, OUTPUT);
  pinMode(rdataPin, OUTPUT);
  pinMode(gdataPin, OUTPUT);
}

byte rbitmap[][8] = {  //red characters
  {0, 0,0,0,0,0,0,0},
  {0, 10, 31, 31, 31, 14, 14,  4},  //heart
  {0, 9, 9, 9, 9, 9, 9, 6}, //U
};
byte gbitmap[][8] = { //green characters
  {0, 14, 04, 04, 04, 04, 04, 14}, //I
  {0, 0,0,0,0,0,0,0},
  {0, 9, 9, 9, 9, 9, 9, 6}, //U
};
void loop(){
  // select the row 0 = top row
  for (int row = 0; row < 8; row++) {
    // send serial data to display 3 = number of characters
    // send the rightmost character out first
    for (int character = 3; character > -1; character--) {
      // select the bit to load into the shift register
      for (int shiftbit = 0; shiftbit < 5; shiftbit++){
        digitalWrite(rdataPin, HIGH);
        digitalWrite(gdataPin, HIGH);
        if bitRead(gbitmap[character][row],shiftbit)
digitalWrite(gdataPin, LOW);
```

```
      if bitRead(rbitmap[character][row],shiftbit)
digitalWrite(rdataPin, LOW);
      // advance the shift register clock
      digitalWrite(clockPin, LOW);
      digitalWrite(clockPin, HIGH);
    }
  }
  // turn off the display
  digitalWrite(rowPin1, LOW);
  digitalWrite(rowPin2, LOW);
  digitalWrite(rowPin3, LOW);
  digitalWrite(rowPin4, LOW);
  digitalWrite(rowPin5, LOW);
  digitalWrite(rowPin6, LOW);
  digitalWrite(rowPin7, LOW);
  digitalWrite(rowPin8, LOW);
  //latch the data
  digitalWrite(rclockPin, LOW);
  digitalWrite(rclockPin, HIGH);
  // set up row selection and turn display back on
  if (row==0) digitalWrite (rowPin1, HIGH);
  if (row==1) digitalWrite (rowPin2, HIGH);
  if (row==2) digitalWrite (rowPin3, HIGH);
  if (row==3) digitalWrite (rowPin4, HIGH);
  if (row==4) digitalWrite (rowPin5, HIGH);
  if (row==5) digitalWrite (rowPin6, HIGH);
  if (row==6) digitalWrite (rowPin7, HIGH);
  if (row==7) digitalWrite (rowPin8, HIGH);

  // Wait to see what we sent to the display ;
  delayMicroseconds(500);
}}
```

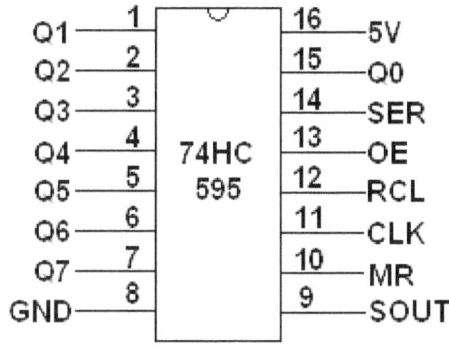

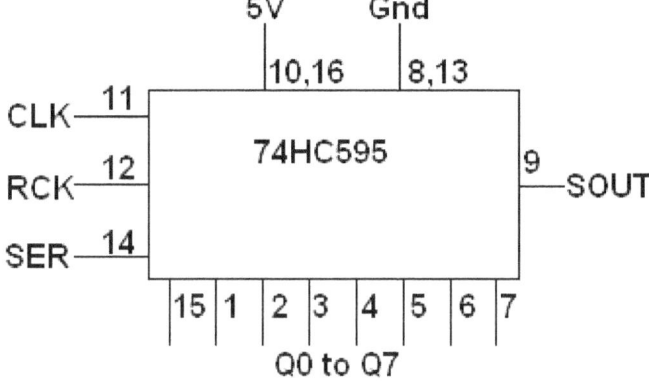

Here is a list of what the 74HC595 pin names mean:

Q0 to Q7 are the shift register output pins.
GND is ground.
SOUT is Serial Output to another shift register if used.
MR is Master Reset when grounded input.
CLK is the Clock to advance the shift register input.
RCL is Register Clock to latch the contents of the shift register.
OE is the Output Enable input, it is active when low.
SER is Serial input.
5V is the power source.

Below there is a picture of what it might look like all wired up. Notice that I used a ribbon cable for the rows. The resistors should have had some insulation on their leads to keep them from shorting.

Chapter 16

Three Color Eight by Eight

LED Array

Recently I was able to get my hands on a three color, as in red, green and blue, eight columns by eight row LED array. The model number of this LED array is CRM-2388ARGB-L. It is very popular on eBay. The LED array did not come with any wiring instructions, so I set down with a nine volt battery and a 1K resistor and figured out how it is wired up.

Underneath it there are two rows of 16 pins each. The top row does not have the part number near it. The bottom row, left most pin, is the closest pin to the part number. There is also a number "1" located near this pin. This is what I found the pin out to be.

```
-----Rows------  ---------------Green-----------------  ------Rows-----
R8  R7  R6  R5  G1  G2  G3  G4  G5  G6  G7  G8  R4  R3  R2  R1
32  31  30  29  28  27  26  25  24  23  22  21  20  19  18  17
-------------------------------------------------------------------
     0     0     0     0      0      0      0      0
     0     0     0     0      0      0      0      0
     0     0     0     0      0      0      0      0
     0     0     0     0      0      0      0      0
     0     0     0     0      0      0      0      0
     0     0     0     0      0      0      0      0
     0     0     0     0      0      0      0      0
     0     0     0     0      0      0      0      0
-------------------------------------------------------------------
1    2    3    4    5    6    7    8    9   10   11   12   13   14   15   16
B1   B2   B3   B4   B5   B6   B7   B8   R1   R2   R3   R4   R5   R6   R7   R8
----------------Blue------------------  -----------------Red-----------------
```

Next I wired up three 74HC595 shift registers with eight 100 ohm resistors each to the red, green and blue array pins. I was reusing the design that had previously worked for a two color eight by eight led array.

For the rows there should have been some row driver transistors, but I directly connected the rows to pins D1-D8 of the Arduino and it worked. You will get a little more brightness with some NPN row driver transistors. However it is already very bright.

Here is the schematic diagram. I chopped off the Arduino on the left to fit the schematic in at full size.

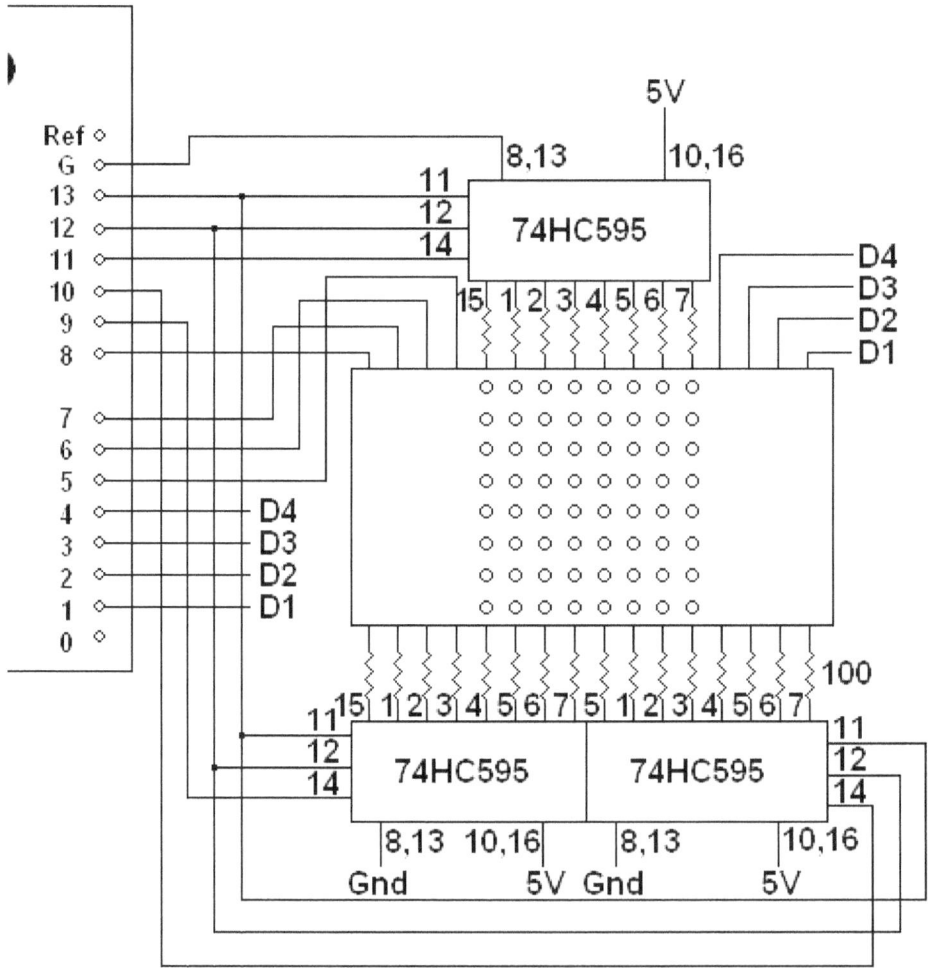

Here is what it looks like all wired up and displaying a smiley face. It is hard to tell what it looks like in black and white. In color, it is a blue face, with a red mouth and eyes. The hair at the top is white.

Next I wrote a quick sketch to output data to the three shift registers and to select the rows one at a time.

```
//*******************************************/
//  Name    : LED 8x8 RGB SR array Driver
//  Author  : Bob Davis
//  Date     : 2 January, 2014
//  Version : 1.0
//*******************************************/
// Assign Pins for the shift registers and Rows
int Row1 = 1;
int Row2 = 2;
int Row3 = 3;
int Row4 = 4;
int Row5 = 5;
int Row6 = 6;
int Row7 = 7;
int Row8 = 8;
int RSerial = 9;
```

```
int BSerial = 10;
int GSerial = 11;
int RCLK = 12;
int SCLK = 13;
// Set up arrays, select one of these three things to display

// RGB Flag
//byte Bbitmap [8]={240, 240, 240, 240, 0, 0, 0, 0};
//byte Gbitmap [8]={15, 0, 15, 0, 255, 0, 255, 0};
//byte Rbitmap [8]={0, 15, 0, 15, 0, 255, 0, 255};
// Blue Face
byte Bbitmap [8]={60, 126, 255, 219, 255, 219, 102, 60};
byte Gbitmap [8]={60, 126, 129, 165, 129, 129,   0,  0};
byte Rbitmap [8]={60, 126,   0,  36,   0,  36,  24,  0};
// Red Car
//byte Gbitmap [8]={8, 0, 0, 0, 0, 0, 255, 255};
//byte Bbitmap [8]={255, 255, 255, 131, 0, 0, 0, 0};
//byte Rbitmap [8]={8, 0, 0, 124, 255, 189, 0, 0};
// Set the pins to output to the array
void setup() {
  pinMode(Row1, OUTPUT);
  pinMode(Row2, OUTPUT);
  pinMode(Row3, OUTPUT);
  pinMode(Row4, OUTPUT);
  pinMode(Row5, OUTPUT);
  pinMode(Row6, OUTPUT);
  pinMode(Row7, OUTPUT);
  pinMode(Row8, OUTPUT);
  pinMode(BSerial, OUTPUT);
  pinMode(GSerial, OUTPUT);
  pinMode(RSerial, OUTPUT);
  pinMode(RCLK, OUTPUT);
  pinMode(SCLK, OUTPUT);
}
void loop() {
  for (int row = 0; row <8; row++){
    //turn everything off
    digitalWrite (Row1, LOW);
    digitalWrite (Row2, LOW);
    digitalWrite (Row3, LOW);
    digitalWrite (Row4, LOW);
```

```
  digitalWrite (Row5, LOW);
  digitalWrite (Row6, LOW);
  digitalWrite (Row7, LOW);
  digitalWrite (Row8, LOW);
  // set up the data in the shift registers
  for (int shiftbit = 0; shiftbit <8; shiftbit++){
    digitalWrite(BSerial, HIGH);
    digitalWrite(GSerial, HIGH);
    digitalWrite(RSerial, HIGH);
    if bitRead(Bbitmap[row],shiftbit) digitalWrite(BSerial, LOW);
    if bitRead(Gbitmap[row],shiftbit) digitalWrite(GSerial, LOW);
    if bitRead(Rbitmap[row],shiftbit) digitalWrite(RSerial, LOW);
    digitalWrite(SCLK, LOW);
    digitalWrite(SCLK, HIGH);
    }
  //latch the data
  digitalWrite(RCLK, HIGH);
  digitalWrite(RCLK, LOW);
  // turn on the associated row
  if (row == 0) digitalWrite (Row1, HIGH);
  if (row == 1) digitalWrite (Row2, HIGH);
  if (row == 2) digitalWrite (Row3, HIGH);
  if (row == 3) digitalWrite (Row4, HIGH);
  if (row == 4) digitalWrite (Row5, HIGH);
  if (row == 5) digitalWrite (Row6, HIGH);
  if (row == 6) digitalWrite (Row7, HIGH);
  if (row == 7) digitalWrite (Row8, HIGH);
  delay(2);
} }
```

Once I got everything working, I played around with the arrays to make it display different things. I started with the smiley face because I had already designed one for a dual color eight by eight array. To create your own array, start with the first number, it is the top row. Each number in the array moves down one row in the display. The numbers values are 128, 64, 32, 16, 8, 4, 2, and 1. To light up the first and last LED you would just add 128+1 to get 129. To light the middle two LED's you would add 16+8 to get 24. Using this pattern you can create your own arrays or make mine look even better.

Chapter 17

Dual Color 8 by 40 LED Array

You can build your own LED sign from scratch, or you can use an already assembled and tested module such as the Cadaces CDVO-1 that can be found in old Signature Electronic Signs. This module from Cadaces has eight five by eight dual color LED arrays for a total of 40 by 8 LED's. There are TPIC6B595 drivers for the columns and a 74138 selects the rows via some high power transistor drivers.

Each Cadaces module can consume over two amps at five volts when all of the LED's are lit. However when you are just displaying text it is possible to run one of these modules directly off the five volts from the USB jack of the computer. For more power, there are small five volt 2.5 amp AC adapters available that will usually run two of these displays.

The Cadaces modules have a 10 pin connector that connects to the Arduino. A second 20 pin connector is used to daisy chain up to four of these modules together in series. The column signals include red serial data, green serial data, data clock, and data latch. The row signals are data zero, data one, data two, and row enable. When row enable is off the entire sign is off, it is used to blank the display while updating the columns. The other two pins are ground.

Here is how the 10 pin connector is wired up to the Arduino. The schematic is shown form above as though you were looking down through the adapter. You will need a five pin header to connect one row of the ribbon cable to the Arduino. The other row is connected via four jumper wires.

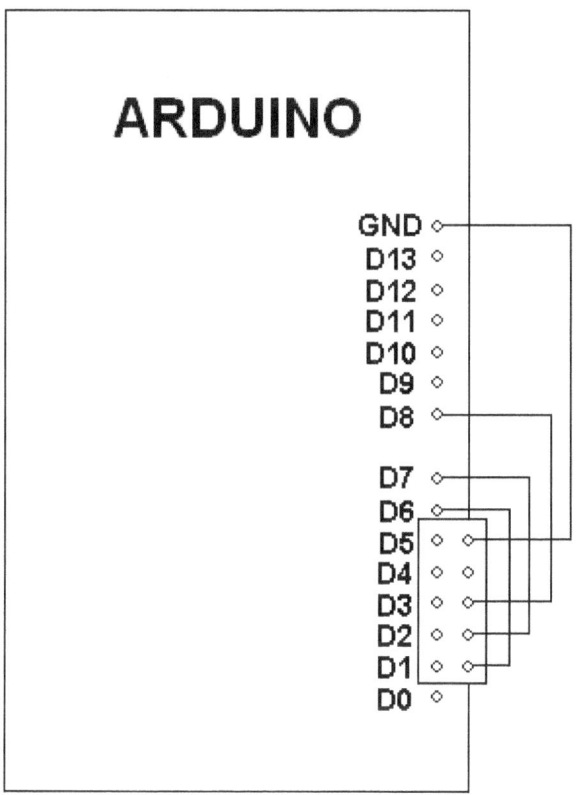

Here is a picture showing how to connect the ten pin connector to an Arduino. The green wire going off the left side of the picture is five volts power going to the red power wire to power up the LED sign.

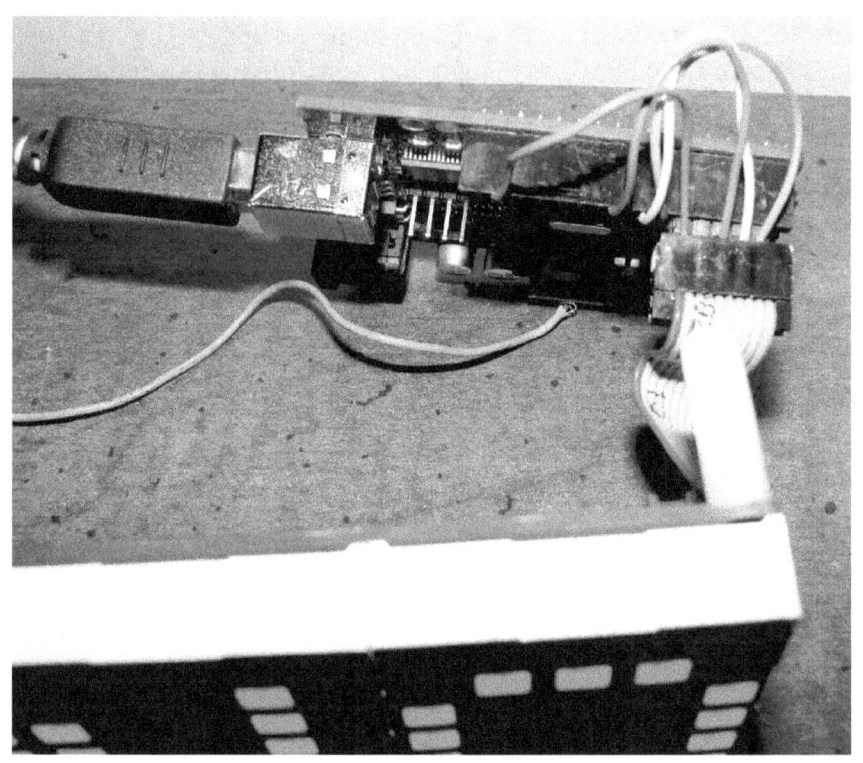

This is a picture of the ten pin connector on the Cadaces sign.

Let me translate some of the signal names:

TPIC6B595 Pins
Pin 3 - R Ser is the Red Serial Data input.
Pin 3 - G Ser is the Green Serial Data input.
Pin 13 - Clk is the Serial Clock input.
Pin 12 - RClk is Register Clock, or "Latch the Data".
GND is Ground.

74HCT138 Pins
Pin 4 - REN is Row Enable, or "Turn the display on".
Pin 3 - R3 is the third bit of the row selector.
Pin 2 - R2 is the second bit of the row selector.
Pin 1 - R1 is the first bit of the row selector.

You will need the following parts to make this project.
1. Arduino adapter or 5 pin header and 4 jumper wires.
2. One or two Cadaces modules.
3. One or two ribbon cables.
4. One Five volt 2.5 amp AC adapter.

Here is what a home made adapter looks like from below. That is not the neatest wiring job in the world but it does get the job done. This is a model that used an Arduino shield.

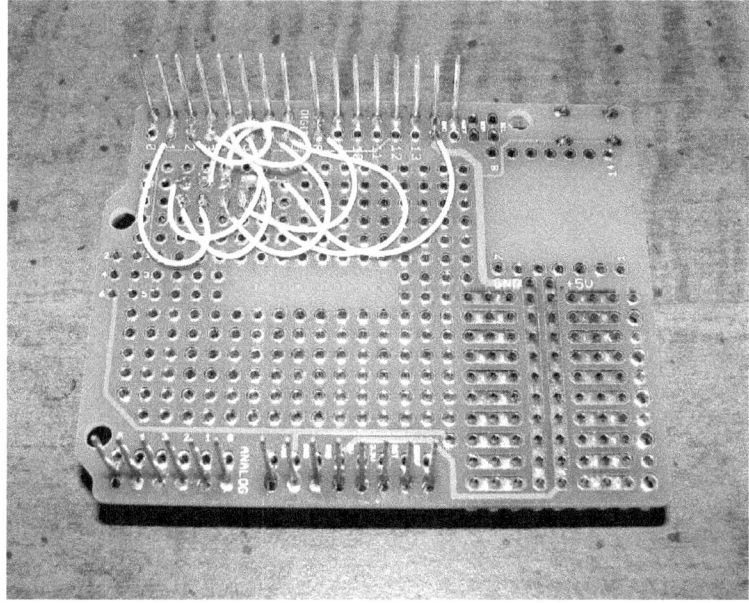

Here is the Arduino shield adapter as it is seen from above:

Here is the code or sketch for a dual color stationary sign, it is written for two Cadaces modules that are wired up in series via ribbon cables.

```
//*********************************
//  Name    : Cadaces Driver
//  Author  : Bob Davis
//  Date    : 23 February, 2013
//  Version : 1.0
//*********************************
// Pins for the row drivers
int row1Pin = 2;      //R1
int row2Pin = 7;      //R2
int row3Pin = 1;      //R3
int rowEnable = 8;    //REN
int rclockPin = 6;    //RCK
int clockPin = 3;     //SCK
int dataPin = 5;      //RSD
int gdataPin = 4;     //GSD
// Set the pins to output to the sign
void setup() {
```

```
    pinMode(row1Pin, OUTPUT);
    pinMode(row2Pin, OUTPUT);
    pinMode(row3Pin, OUTPUT);
    pinMode(rowEnable, OUTPUT);
    pinMode(rclockPin, OUTPUT);
    pinMode(clockPin, OUTPUT);
    pinMode(dataPin, OUTPUT);
    pinMode(gdataPin, OUTPUT);
}
//=== Character Array ===
// Characters are A, B, C, etc.  Only upper case, no symbols.
byte alphabets[][8] = {
  {0, 04, 10, 17, 17, 31, 17, 17}, //A
  {0, 30, 17, 17, 30, 17, 17, 30}, //B
  {0, 14, 17, 16, 16, 16, 17, 14}, //C
  {0, 28, 18, 17, 17, 17, 18, 28}, //D
  {0, 31, 16, 16, 31, 16, 16, 31}, //E
  {0, 31, 16, 16, 31, 16, 16, 16}, //F
  {0, 14, 17, 16, 16, 19, 17, 14}, //G
  {0, 17, 17, 17, 31, 17, 17, 17}, //H
  {0, 14, 04, 04, 04, 04, 04, 14}, //I
  {0, 07, 02, 02, 02, 02, 10, 14}, //J
  {0, 17, 18, 20, 24, 20, 18, 17}, //K
  {0, 16, 16, 16, 16, 16, 16, 31}, //L
  {0, 10, 21, 21, 21, 17, 17, 17}, //M
  {0, 17, 25, 25, 21, 19, 19, 17}, //N
  {0, 14, 17, 17, 17, 17, 17, 14}, //O
  {0, 30, 17, 17, 30, 16, 16, 16}, //P
  {0, 14, 17, 17, 17, 17, 19, 15}, //Q
  {0, 30, 17, 17, 30, 20, 18, 17}, //R
  {0, 14, 17, 16, 14, 01, 17, 14}, //S
  {0, 31, 04, 04, 04, 04, 04, 04}, //T
  {0, 17, 17, 17, 17, 17, 17, 14}, //U
  {0, 17, 17, 17, 10, 10, 10, 04}, //V
  {0, 17, 17, 17, 21, 21, 21, 10}, //W
  {0, 17, 17, 10, 04, 10, 17, 17}, //X
  {0, 17, 10, 10, 04, 04, 04, 04}, //Y
  {0, 31, 8, 04, 02, 04, 8, 31}, //Z
  {0, 0, 21, 14, 31, 14, 21, 0}, //*
};
byte numbers[][8] = {
```

```
    {0, 04, 12, 04, 04, 04, 04, 14}, //1
    {0, 14, 17, 01, 02, 04, 8, 31}, //2 note that 08 is illegal.
    {0, 14, 01, 01, 06, 01, 01, 14}, //3
    {0, 17, 17, 17, 31, 01, 01, 01}, //4
    {0, 31, 16, 16, 14, 01, 17, 14}, //5
    {0, 14, 16, 16, 30, 17, 17, 14}, //6
    {0, 31, 01, 01, 02, 04, 8, 16}, //7
    {0, 14, 17, 17, 14, 17, 17, 14}, //8
    {0, 14, 17, 17, 15, 01, 01, 01}, //9
    {0, 14, 17, 17, 17, 17, 17, 14}, //0
}
byte bitmap[][8] = {  //red characters
    {0, 0,0,0,0,0,0,0},
    {0,  4, 10, 17, 17, 31, 17, 17}, //A
    {0, 30, 17, 17, 30, 20, 18, 17}, //R
    {0, 28, 18, 17, 17, 17, 18, 28}, //D
    {0, 0,0,0,0,0,0,0},
    {0, 14, 04, 04, 04, 04, 04, 14}, //I
    {0, 17, 25, 25, 21, 19, 19, 17}, //N
    {0, 14, 17, 17, 17, 17, 17, 14}, //O
    {0, 0,0,0,0,0,0,0},
    {0, 14, 17, 16, 14, 01, 17, 14}, //S
    {0, 0,0,0,0,0,0,0},
    {0, 14, 17, 16, 16, 19, 17, 14}, //G
    {0, 17, 25, 25, 21, 19, 19, 17}, //N
    {0, 0,0,0,0,0,0,0},
}
byte gbitmap[][8] = {  //green characters
    {0, 0,0,0,0,0,0,0},
    {0,  4, 10, 17, 17, 31, 17, 17}, //A
    {0, 30, 17, 17, 30, 20, 18, 17}, //R
    {0, 0,0,0,0,0,0,0},
    {0, 17, 17, 17, 17, 17, 17, 14}, //U
    {0, 14, 04, 04, 04, 04, 04, 14}, //I
    {0, 0,0,0,0,0,0,0},
    {0, 14, 17, 17, 17, 17, 17, 14}, //O
    {0, 0,0,0,0,0,0,0},
    {0, 14, 17, 16, 14, 01, 17, 14}, //S
    {0, 14, 04, 04, 04, 04, 04, 14}, //I
    {0, 0,0,0,0,0,0,0},
    {0, 17, 25, 25, 21, 19, 19, 17}, //N
```

```
  {0, 0,0,0,0,0,0,0},
}
void RunSign() {
  for (int row = 7; row > 0; row--)  {
    // turn off display
    digitalWrite(rowEnable, HIGH);
    digitalWrite(rclockPin, LOW);
    // send serial data to display 14 = number of characters
    for (int character = 0; character < 14; character++)
    {
      for (int shiftbit = 5; shiftbit > -1; shiftbit--) {
    digitalWrite(gdataPin, LOW);
    digitalWrite(dataPin, LOW);
    if bitRead(gbitmap[character][row],shiftbit) digitalWrite(gdataPin,
HIGH);
    if bitRead(bitmap[character][row],shiftbit) digitalWrite(dataPin,
HIGH);
    digitalWrite(clockPin, HIGH);
    digitalWrite(clockPin, LOW);
      }
    }
    //latch the data
    digitalWrite(rclockPin, HIGH);
    // set up 74138 row sesection and turn display back on
    if bitRead(row,0) digitalWrite (row1Pin, HIGH);
        else digitalWrite(row1Pin, LOW);
    if bitRead(row,1) digitalWrite (row2Pin, HIGH);
        else digitalWrite(row2Pin, LOW);
    if bitRead(row,2) digitalWrite (row3Pin, HIGH);
        else digitalWrite(row3Pin, LOW);
    digitalWrite(rowEnable, LOW);
    // Wait to see what we sent to the display
    delayMicroseconds(500);
  }
}
//=== L O O P ===
void loop() {
  RunSign();
}
```

Bibliography

Programming Arduino
Getting Started With Sketches
By Simon Mark
Copyright 2012 by the McGraw-Hill Companies

This book gives a thorough explanation of the programming code for the Arduino.
However the projects in the book are very basic. It does cover LCD's and Ethernet adapters.

Getting Started with Arduino
By Massimo Banzi
Copyright 2011 Massimo Banzi

This author is a co-founder of the Arduino.
This book has a quick reference to the programming code and some simple projects.

www.ingramcontent.com/pod-product-compliance
Lightning Source LLC
Chambersburg PA
CBHW081547170526
45166CB00009B/2605